Vibrational Matter

Book 1 in the Vibrational Matter Series

By Steve Preston

2nd Edition

Table of Contents

A New Theory

This book is about how matter and energy are made in our universe. No; they are not made with length, width, and height as you were told everything in the universe was made. Instead, everything in the universe is made of vibrations. Unfortunately, it isn't that simple as the things that are vibrating to make matter and energy are both invisible. Matter is made up of something now called Aether while energy is made of something called Electricity. This is the first of a series that will, hopefully allow you to understand our world in a much different way so you have a chance at understanding what the universe is, what matter and energy and even life is. We will also begin to understand what time really is. I will not be providing massive equations and stating "it can be determined" after some computation that makes no sense, but let me tell you up front there will be a level of confusion if you have not viewed the universe before as being made with vibrational matter.

If you have heard about the unified particle theory, this is not about that, but it did put a lot of Physicists into discovering this universal fact. Einstein spent the last half of his life searching for a particle that would be the building block of everything. What we know today is that atoms are made up of something called Baryons like "electrons" and "protons". These baryons are not the building blocks of matter because scientists found things that make up the baryons called Bosons. These things were the smallest particles that could make up matter or could exist as an entity to be sure, but there were several types. What a dilemma. Einstein had spent his whole life searching for something that could not be. Soon scientists found out that something called a fermion was less than a particle. Oh! My! This must be the smallest particle of matter. A graviton, for instance, is not quite a particle. It doesn't conform to matter in that it displays gravity, but there is no mass to make the gravity that is sensed. Unfortunately, several types of fermions were discovered, in fact photons themselves with their unusual character of sometimes having mass and sometimes not having mass put them into the fermion class.

What in the world are we to do????? [I'm making tiny little cry sounds. The cry sounds are fake so don't worry about me.]

Eureka!

The answer is both simple and almost impossible to understand. Everything---I mean everything, is made of "a special nothing we can call Aether" that is vibrating. As This Aether doesn't exactly exist, the term used is quantum fluctuations. These quantum fluctuations make up all matter, all electro-magnetics, all nuclear energy, all photons, and even all lifeforces. This book is about the thing that is smaller than the unified particle and more basic even that a fermionic sub-particle. In order to study it, we need to convert the universe into a vibrational concept. When that happens, we find several things. First, we find that conversion of and control of matter is no longer a mystery, it is reasonable and the concept of heaven is no longer impossible. Heaven now becomes necessary. The concept of transference to another universe, say heaven, is no longer some figment of the imagination of a religious leader. It now becomes fact. This is a concept that will help you understand yourself, your religion, your life and death, and your entire universe.

Just to reiterate what I'm trying to say and show you in this book is that the smallest component of matter isn't matter at all. I know it sounds absurd and you are thinking you picked up a book by a nutcase, but, hopefully, you will soon see why our world is made up of vibrational matter.

Electricity Example

To understand matter, you need to understand electricity. You know about electricity, but it doesn't exist. The definition of Electricity is the <u>POTENTIAL</u> to provide Electro-magnetic work. Just like any potential energy it doesn't exist but we use it every day by adding something called Magnetism. The faster this combination of electricity and magnetism vibrate the more Energy their combination supplies our environment.

- *Low frequencies are called Radio Waves*

- *Higher frequencies become visible light*

- *Still higher becomes X-Rays that can shoot through your body.*

- *Still higher this same electro-magnetic combination become gamma wave that can kill you*

Think of matter as the same thing except we use something called Aether as the POTENTIAL to provide an Aethereo-Gravitational construct which we like to call matter. Just like any "potential" it doesn't exist but we use it every day by adding something called Gravity. The faster this combination of Aether and Gravity vibrate the more density their combination supplies our environment.

- *Low frequencies start making Fermions [quasi-mass]*

- *Higher frequencies make Electrons*
- *Still higher forms Hydrogen*
- *Higher still make gold*
- *Soon Matter becomes what is known as a black hole when it vibrates so fast that it becomes almost total Gravity. Its no longer sensed as matter.*

Therefore, if you can just think about matter as being the same as electricity, you might not read the book. If you are still confused about how atoms get huge and all the weirdness of matter, you may still want to continue.

Not About Rome

To be honest with you, this book was going to be a book about Rome, but I got too confused. The place was so weird I changed the topic to something a little less bizarre like this vibrating matter stuff.

Cleanliness Wasn't Easy

Romans, for instant, didn't wash themselves. Instead, they scrapped themselves after smearing oil on their bodies to get clean at their common bath areas. Some of the bath scrapper things are shown above. The Romans were all about community. Their public toilets had dozens of seats along all 4 walls with keyhole slots in the lids. One of these community areas is shown to the right. I suppose it was easy to pass the paper or "sponge on a stick" as was used by these highly social people. While it seems somewhat inappropriate now, at the end of the day, the urine was collected and used as a component of their toothpaste and as a laundry ingredient to keep those toga things white and wearable. No;

I don't know if the toothpaste was named PEEPASTE nor do I know if they ran adds about how yellow your teeth could become. By the way, these bathrooms had to have one wall removed, especially if all 36 locations were in use at the same time.

Eating Was Gross

When these civilized people went to eat, some of the food that was served included meat stuffed in a cow utter or sheep bladder. Flamingo tongue [a whole lot of them were required to fill you], baked mice, badger ears, and wolf nipples. Gargum was another delicacy. To make this yummy soup, entrails of one large fish and several smaller ones were placed in a container, salted, and then exposed to the sun until the fish had gone putrid. The liquid was drained off and used as the basis for a relish on just about anything.

If a stomachache arose, don't worry. The simple cure was to wash your feet and drink the wash water. Whether a stomachache was gained or not, overeating was encouraged and almost demanded. A feather was provided for those wishing to throw up a little to increase the eating capacity.

Their Leaders Were Nuts

One of their leaders, named Nero, had a quaint custom. At a dinner party he tied Christians to poles, soaked them in tar and lit up the area for a little ambiance. Another emperor named Commode changed the name of Rome to Commodia and made everyone call him god. Who in the world wants a god named after a toilet? Emperor Caligula was also a fruitcake. His sister, Drusilla died and he was very sad. Therefore, he proclaimed that no one could laugh or take a bath for a year. The penalty for transgression was death, if you could get close to the laughing guy.

The People Were Weird

Prostitutes were easy to spot in Rome. They were required to die their hair blond or wear a blond wig at all times. Luckily there were no Romans that lived in California. It is so confusing out there.

Just like we have today, the Romans had a steroid problem. But in this case, the steroids of ancient Rome were dried boar's dung. Chariot racers often took a drink made from the dung before major events. The custom of the pretty girls kissing the winners at races was not started in Rome, I'm told. Even if they brushed their teeth with some PEE TOOTHPASTE, it didn't help.

There is simply no way I'm writing about Rome. The place was just too bizarre. I need to concentrate on normal things.

This book is about vibrational matter -Not Oddness! I think starting off with Ekpyrotic Membranes is a good way to show it isn't odd.

The word Ekpyrotic simply means explosion, but it is more fun to say.—*Ekpyrotic- Ekpyrotic -Ekpyrotic.*

Ekpyrotic Membrane

The Ekpyrotic /Membrane Theory is a good one. As I tried to say, it indicates that there must have been at least 2 adjoined universes that splatted together to initiate what we call the Big Bang some 15 billion years ago or the Big Bang simply could not have occurred.

This SPLAT action caused everything to emerge from a 'fireball' with a temperature of 10 billion degrees. When the two splatted together, <u>the energy that became matter was introduced</u>. According to this theory, tiny quasi-particles called fermions were all over the place. I'm sorry, but you will have to remember this word, because quite a bit of our universe uses these fermions as the basic unit. It is when that Aether stuff I mentioned before almost becomes something real. I think John Keely invented the word Aether in the 19th century, but it has now been adopted by so pretty big names like Einstein so stick with me here.

When the universes collided 15 billion years ago, the Aether started vibrating making fermions. Fermions got all twisted around and turned into complete particles we call Bosons. It is these bosons that make up most of what we call matter. The question is what actually caused the fermions to become matter and what things are necessary in this universe for sustainment?

The reason the fermions became matter is simple. They began to vibrate. This vibration made them exist in this universe or our linked universe.

It is the vibrational similarity of fermions that make them exist in this world or another. It is the natural allowance for various frequencies in a dimensional string that causes various atoms to form. Vibrations are fairly simple to understand but they can make our world difficult to understand. As I wondered about this whole vibration thing I was stopped by Einstein.

Einstein

This crazy guy was going around saying that the time-dimension energy equation was $E=MC^2$. I knew in my heart that he was wrong. After all, all energy equations are of the same form factor.

- $E = \frac{1}{2} KX^2$ *[universal law of potential energy]*

- $E = \frac{1}{2} LI^2$ *[universal law of magnetic energy]*

- $E = \frac{1}{2} CV^2$ *[universal law of capacitive energy]*

- $E = \frac{1}{2} MV^2$ *[universal kinetic energy equation]*

- $E = \frac{1}{2} I\alpha^2$ *[Universal Inertial Energy equation]*

The Missing Half

Where was the ½? No one seemed to care, but it bothered me just like a woman's ability to talk for many more hours than I could. The more I tried to show Einstein was wrong the more it seemed to be right. It was as if ½ the mass in the universe was invisible and the invisible mass had to be "used in the

13

equation". If you wonder what else is invisible, simply look at photons.

Light Anomaly

This book is not a mathematical listing of numbers that add up to some esoteric characterization of matter that no one can dispute. It is also not a complex discussion that takes a scientist to understand what in the world I'm talking about. This book doesn't, necessarily, standalone. Instead, it is a discussion of discussions and a theory of theories, so to speak. Hopefully, by looking a wide assortment of ideas and bits of evidence it clears up many of the anomalous details presented by other theories of matter.

For instance, the atom is not the essence of matter.

There certainly are atoms or at least atomic characteristics, but basing matter on atoms and molecules and nuclear forces and covalent bonds and all the other things presented to you previously will not allow you to understand more basic elements of life. All matter should be thought of as being made up of vibrating fields. This fundamental characteristic will change how you look the entire world. I know you think you are comfortable with the way things are, but let me test you on that.

Light Example

Let me take a simple example- "LIGHT". I briefly described this earlier, but what do you know about this thing? If you said NOTHING! You would be almost completely correct.

No, no, no my friend. Light is sometimes a particle and sometimes it acts like there is no mass whatsoever, but there is an electromagnetic wave or frequency associated with the color of the light [Whatever that is!]. We know that the faster the photon of light vibrates the more powerful it becomes. Soon the fast vibrating photon becomes dangerous to humans as it can go right through the body [x-ray] and if it slows down too much it changes into something we call radio waves. I know you are thinking that these radio waves must not exist because they don't produce light and they have no mass, but let me assure you that sometimes these photon things do act like normal matter. In fact, normal matter doesn't act like normal matter all the time. That's where the book comes in. We need to redefine matter so that this will all make sense. If you look at the diagram below, there is a wiggly line. The faster wiggling represents a prime particle vibrating faster and faster. Radio waves turn into light that turns into the deadly gamma rays.

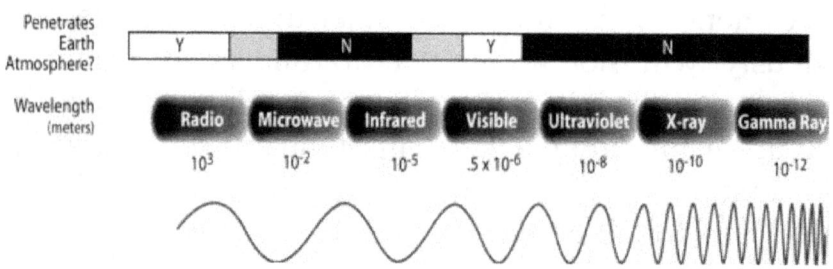

Today the chart continues even farther. The vibrational frequencies are in the exahertz range [exahertz mean "quintillion cycles per second"] in the chart below so don't go out and try to whistle some gold in your pocket. It doesn't work like that but do keep that exahertz thing in the back of your mind right now, because it just sounds weird right now and I want to bring you up to speed so you can see a better picture of reality.

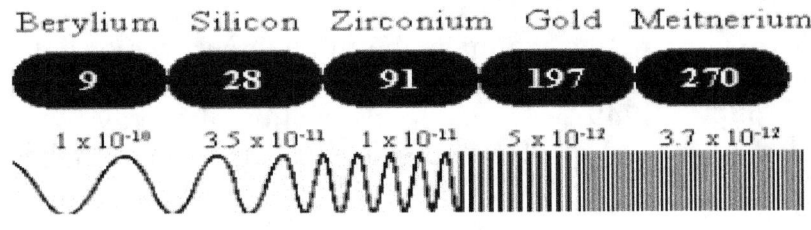

History of the Atom

A long time ago, the Greeks decided that there was something called an atom that was the basic unit or all matter. The idea sort of stuck so everyone went about trying to prove what he or she wanted "matter" to be. All the greats started building on the story and it was a good one----for a while. A general timeline is shown next.

Inventor	Date	Description
Democritus	440 BC	He proposed the concept of an atom to describe the indivisible and indestructible particles that were thought to compose the substance of all things.
Leonardo da Vinci	1500	He pointed out that animals could not survive in an atmosphere that could not support combustion.
Copernicus	1543	He reinvented the heliocentric model of the universe.
William Gilbert	1544-1603	He investigated magnetism & static electricity.
Otto von Guericke	1602-1686	He invented first electrostatic generator
Robert Boyle	1662	He found that the volume occupied by the same sample of any gas at constant temperature is inversely proportional to the pressure.

Isaac Newton	1687	He published his "Philosophiae Naturalis Principia Matematica". He set the foundations of mechanics, gravitation, a theory of light, and calculus.
Francis Hauksbee	1666-1713	He studied electrostatic repulsion and made the first mercury vapor light
Emilie du Chatelet	1706-1749	He studied infrared emission & measured conservation of energy
Benjamin Franklin	1706 - 1790	He studied electrical charges & labeled them "positive" & "negative", and a whole lot more!
Luigi Galvani	1737-1798	He discovered electrical effects between different metals, and in biological cells
James Watt	1760's	His claim to fame is that he improved on the Newcomen Engine.
Charles de Coulomb	1785	He measured the attractive and repulsive forces of electrically charged particles, and discovered that these forces are inversely proportional to the square of the distance.
Jacques Charles	1787	He studied the volume changes of gases with changes in temperature.
Alessandro Volta	1745-1827	He developed the forerunner of the electric battery.
André Marie Ampère	1775-1836	He established a measurable relationship between electricity and magnetism.
Hans Christian Oersted	1777-1851	He developed experiments in electromagnetism.
Carl Friedrich Gauss	1777 - 1855	He experimented with electrical charges and magnetism, and established a method for measuring magnetic fields.
Georg Simon Ohm	1789 - 1854	As you would expect, his work led to the mathematical relationship between voltage, current, and resistance called "Ohm's Law".
Michael Faraday	1791-1867	He developed measurement methods in capacitance & electromotive force.
Sir Humphrey Davy	1809	Discovered many metals [sodium, potassium etc,] and invented the first electric light bulb.

Amadeo Avogadro	1811	He demonstrated that equal volumes of all gases under the same temperature and pressure contain the same number of molecules, and that a fixed number of molecules of any gas will weigh proportional to its molecular weight. Presently the accepted value for the Avogadro number is 6.023 x 10^23 molecules per gram-mol.
Joseph Henry	1797 - 1878	He discovered electromagnetic self-inductance, and invented the electromagnetic relay which- le to the telegraph.
Sadi Carnot	1824	He published his Reflexions sur la Puissance Motrice du Feu, setting various outstanding principles that constitute the basis of actual Thermodynamics.
Friederich Wöhler	1828	He synthesized the first organic compound from inorganic compounds, preparing Urea by reacting lead cyanate with ammonia. It was a nasty job.
Robert Brown	1828	He first described Brownian motion. Before him it may have been called Bobian motion.
James Prescott Joule	1818 - 1889	He developed more theories on conservation of energy, thermodynamics, and resistance heating
Michael Faraday	1831	He showed the relation between magnetism and electricity is dynamic. He showed that not only was magnetism equivalent to electricity in motion but also, conversely, electricity was magnetism in motion. Later, Clerk Maxwell summarized in concise form the electromagnetic theory.
Jöns Jacob Berzelius	1835	He demonstrated that the hydrolysis of starch is catalyzed more efficiently by malt diastase than by sulfuric acid. He published the first general theory of chemical catalysis.
Julius Robert Mayer	1842	He enunciated the Law of Conservation of Energy-1st Law of Thermodynamics, after establishing the work equivalent of Heat.

Helmoltz and Mayer	1845	He formulated the Laws of Thermodynamics.
Friederich von Stradonitz	1858	He proposed that carbon atoms can form chains.
Canizzaro	1860	He presented new methods determine atomic weights; Oxygen weight of 16 was adopted as measuring basis of element weights, thus setting Hydrogen's weight, the lightest known element, to approximately 1.
John Reis	1861	He invented the telephone.
Dmitri Mendelejeff	1869	He published a chemical elements arrangement table. This is the basis of the well-known periodic table.
Sir William Thompson	1824 - 1907	This guy was also known as Lord Kelvin, studied thermodynamics & proposed an absolute temperature scale that we use today.
James Clerk Maxwell	1831 - 1879	He developed a set of differential equations known as "Maxwell's Equations", which describe electromagnetic radiation and its interaction with matter.
Nathan Stubblefield	1892	He invented the ground battery, the first radio and the first mobile phone.
Alexander Graham Bell	1847 - 1922	Alexander was one of several inventors of a new telephone similar to John Reis version. He also invented the photophone which transmitted sound over light waves.
Thomas A. Edison	1847 - 1931	With 1,097 U.S. patents in his name, Thomas Edison is considered one of the most prolific inventors in history. His inventions included the stock ticker, phonograph, motion picture camera, the kinetoscope, and the incandescent lighting system.
John Keely	1871	Discovered the etheric force that controls the atomic constitution of matter and invented devices run on harmonic motion.

Josiah Willard Gibbs	1878	He developed the theory of Chemical Thermodynamics introducing fundamental equations and relations to calculate multiphase equilibrium, the phase rule, and the free energy concept.
Wilhelm Konrad Roentgen	1895	This German physicist discovered a new kind of radiation working with the vacuum tube discharge. This radiation was called X-rays.
Max Plank	1899	He introduced the concept that light and all other kinds of electromagnetic radiation, which were considered as continuous trains of waves, actually consist of individual energy packages with well defined amounts of energy quanta, proportional to its vibration frequency.
Nikola Tesla	1856 - 1943	Nikola developed an alternating current system of generators, motors, and transmission lines. Also he worked with a whole lot more that may have been the beginnings of the understanding of vibrational matter.
Heinrich Hertz	1857 - 1894	He described the basic unit of frequency and also made discoveries in electromagnetic transmission, the photoelectric effect. He also developed the spark-gap transmitter and dipole antenna.
Lee de Forest	1873- 1961	He invented the audio vacuum tube which amplified weak electric signals
Guglielmo Marconi	1874- 1937	He built upon earlier ideas and patents to develop a widely used radio system
Philo Farnsworth	1906- 1971	He is an American inventor considered to be the father of the television system
Einstein	1905	He introduced the Special Theory of Relativity, established the Law of Mass-Energy Equivalence, created the Brownian Theory of Motion, and formulated the Photon Theory of Light.
Sir Ernest Rutherford	1911	He proposed his theory concerning the atomic nucleus.
Wilson	1912	He described a cloud chamber that allowed the detection of protons and electrons.

Niels Bohr	1913	He proposed his "solar system" model of the atom.
Ed Leedskalinin	1935	He discovered a way to levitate heavy blocks using some type of magnetic device.
William Shockley	1910-1989	He co-invented the solid-state transistor with John Bardeen and Walter Brattain.
Jack Kilby	1923 - 2005	He patented the first integrated circuit while at Texas Instruments, and then later patented the portable calculator.
Robert Noyce	1927 - 1990	Robert further developed the integrated circuit to include more transistors on a silicon substrate.
Gordon Moore	born 1929	He co-founded Intel in 1968 & he is known for "Moore's Law" which observes that integrated circuit complexity doubles every 2 years.
Leonard Susskind	1960s	He invented string theory, and with it the discovery of multiple universes was explained along with vibrational differences in dimensional objects, and also the holographic principle.
John Hutchison	1979	He discovered a way to levitate of heavy objects, fuse of dissimilar materials such as metal and wood, while lacking any displacement, create the anomalous heating of metals without burning adjacent material, spontaneous fracturing of metals, change the crystalline structure and physical properties of metals, and even make metal sample disappear.
J. Newcombe Hodges	2002	Determined the Tamashii definition of atomic structure. A unified particle controlled my variations in vibration.
Andrea Rossi	2012	He designed the first commercial concept of a cold fusion electrical generator that converted Nickel into Copper matter.

I have left a few people out right now, but we will get to them as we go along. Just kidding! I'm not going to talk about these guys, but I thought it was important to show the changes over time and this gave me a way to introduce a few of the ones that helped shape vibrational matter. The last 2 identified inventors, for instance, will be very important as we go along as will Ed Leedskainin, Nathan Stubblefield, Nikola Tesla, John Keely and a few other guys that began changing how we viewed matter. I know some of these guys are usually described, but in the vibrational world they are some of the masters. What we are going to find out is that all the "impossible things" you have been told about aren't impossible at all.

Modern Concept

Today we believe that besides the characteristics of time-space, the rest of universe is made up of vibrating nothingness that is split into three 3-dimensional dynamos that characterize what we call existence. I know that sounds stupid right now, but I think it will make more sense as we go along. Each dynamo makes one of the universal characteristics of Matter, Force, and Life. If any one of the dynamos is missing, the universe cannot exist. I know some believe a universe without life can exist, but today theories of Participatory Anthropics and quantum fluctuation both agree that without a cognizant life, reality tears apart.

Unreal Forces

The first Dynamo uses as its base something everyone calls ELECTRICITY. The Electricity stuff is the POTENTIAL for forces that hold our universe together or tear it apart. I know you thought you knew what electricity was because you thought we get some of it out of an electric outlet in the home to power things, but you don't. ELECTRICITY does not exist. As it vibrates from an association with Magnetism, different types of electromagnetic stressors can be recognized in our universe including radio wave, light beams, cosmic rays, and even useable current that powers things. This book is not really about these dimensional stressors needed to build a universe.

Unreal Matter

The second dynamo uses as its base something Einstein called Aether. This Aether stuff is the POTENTIAL for matter. As Aether is vibrated from an association with Gravity, different types of matter can be recognized. If it vibrates too fast, it becomes complete gravity or, what we call a black hole. This book is about this particular aspect of dynamo of our universe.

Unreal Life

The last dynamo that allows our universe to exist has as its base the potential for LIFE. When vibrated by its association with something we call the SOUL, life appears and becomes more complex and more powerful the faster it vibrates. If the entity vibrates fast enough, "it" becomes a "released soul that combines together with others to maintain what we call reality. That is also not going to be discussed in this book, but you can keep it in the back of your mind for now.

To really find out about what matter is, we must start at the beginning. When I say beginning, I mean the VERY beginning.

15 Billion Years Ago

Even though this book is about vibrational matter, the stage must be set. The anomaly of the earth and how it plays must be established, the use of terms must be defined, and the time sequencing must be solidified. Please be patient and you will be rewarded with new knowledge that will help you understand all about what makes sense with respect to matter and, at the same time his place called Heaven that is pushed around in churches will make a whole lot of sense as well. I promise. At least, I promise you will learn something if you really want to.

Everyone has already determined what he believes to be truth. With respect to our beginnings, some believe the Big Bang Theory. Some believe the literal "Creation Story" in 7 days. Some believe Survival of the Fittest and the Evolution Theory. Some believe everything their government tells them. Some believe everything their clergy tells them. Some believe what they feel in their heart. Some believe only what they see. Some people don't believe in anything. Before we can start tying details from science in some logical pattern, we need to see the problems with various concepts. I'm going to start with a big one.

Big Bang Problems

26

"Scientific creation" has problems with its major "creation" theory, which indicates that 15 billion years ago the universe was created, **without help**, in a huge explosion or "big bang". Research into the details of the theory has shown it to be terribly flawed. That doesn't mean that many of the components of the theory have no relevance and that there is another theory that provides all the answers because there is not; at least as far as I can tell. What I want to show here is that people take these theories as TOTAL TRUTH. If the facts don't meet the theory, they figure that the facts must be wrong rather than the other direction. Here is a list of the various components in dispute with regard to this, almost universally accepted, but substantially flawed, conclusion about spontaneous eruption of our universe.

There Is Not Enough Matter

In the universe there is only 1/100th the amount of matter needed to support the Big Bang according to W.H. Mcrea, "Science and Creation", 1971.

Fast Stellar Rotation

The Stellar rotation is too fast to support the big bang theory according to Richard Johnson, "No Way Out", 1963.

There Is Not Nearly Enough Antimatter

If matter was created during the big bang, an equal amount of antimatter must also have been created, and there is almost none according to R.M. Somerville, "Cosmic Mysteries", 1990.

The Universe Is Too Lumpy

The Big Bang doesn't allow for ripples in the cosmic fabric that created the galaxies and the large-scale structure in the universe. The law of entropy indicates that the universe should seek its most random condition after this long time from the "Big Bang". Instead, lumps of galaxies are all over the place according to P. Peebles, "Science", 1982, and Guth and Slunhart "Scientific American", 1984.

Red Shift Dilemma

Video spectrum red shifts of quasars are too great. These red shifts indicate that we are able to view stars that are almost 15 billion years old and at unbelievable distances away when using the big bang model. The stars would have to have been larger, brighter, and denser than anything that we could possibly imagine. The problem is that **several** of these supposedly immense stars have been located. They all have the same anomaly and everyone is unbelievable, according to "Time-Life", Stars, 1988.

The Background Radiation is Too Low

Too little radiation has been recorded from the stellar masses. This does not support a growing universe model as indicated by the big bang theory- according to William Corless, "Stars, Galaxy, Cosmos", 1982.

No Monopoles

While Electrical charge particles are everywhere, no magnetic charge particles or monopoles have been found in our universe. Presumably, the intense heat of the early Big Bang should have produced monopoles in great amounts just like electrical charges.

Infinite Mass

Without some external force, the Big Bang could only have happened when particles had increased in density to an almost infinite level. No one could explain this requirement and how it could have been satisfied.

As a model, I believe that the Big Bang actually did occur and it is a great way to describe what happened; matter simply didn't get created as indicated in this theory. Something called Ekpyrotic detail brings us closer to the truth.

Say it with me EKPYROTIC, EKPYROTIC. Never mind! This is not a chanting book. It is a serious book about the foundation of our very universe so I will try to stay focused.

I know some of this Physics stuff can get bizarre, but I will try to keep it down so you can actually enjoy the book. Please stay with it and I think you will be glad you did.

As far as a main issue with the Big Bang concept, it is that by red shift calculations, the explosion began on or near earth. Of course, that cannot be and a whole new science came into being called Anthropics. That is too weird for this book so we will go on to the Big Splat.

Big Splat Theory

The Big Splat comes to the rescue. If the Big Bang wasn't enough, this new "Big Splat" theory should have completed our understanding of how matter was created.

It's actually called the Ekpyrotic Universe theory [I love saying that word, but I really have no idea what it means,]

The whole theory is of a time just before the Big Bang. Let me kind of walk you through this Ekpyrotic Theory just a little. The theory is really a subset of something called the M-theory [membrane theory] which identifies at least 12 dimensional components described as strings or membranes [combination of 2 strings reacting as a system.]. We'll get into that in just a bit, because strings and membranes don't have to be confusing. They can be fun and enlightening. They also will get us one step closer to discovering the secrets of vibrational matter.

The Big Splat, essentially, proposes that there is an unseen parallel universe to our universe and before we even had a universe, there were multiple universes.

I know we have all heard about this parallel universe thing, but in one of the generalizations of this theory some 15 billion years ago, before everything supposedly emerged from a 'fireball' with a temperature of 10 billion degrees in 4-dimensional space [called the Big Bang Event], we had only two perfectly aligned, four-dimensional surfaces. One became our universe and the other became a parallel universe, invisible to us. When the two splatted together, the energy that became matter was introduced. According to this theory, tiny particles called fermions were all over the place. These fermions were missing components that they needed to exist in our universe or the other one.

When the universes collided, the fermions turned into complete particles we call Bosons. It is these bosons that make up most of what we call matter.

I know this sounds weird, but it is gaining substantial acceptance from many universities and researchers, so you need to at least know what they are talking about. The Ekpyrotic theory doesn't get us too close to what makes matter, but it does help us begin to see what is happening because these boson things are vibrating like mad. The more they vibrate the larger the effective particle group becomes. The vibrational significance is what vibrational matter is all about.

Why Splat?

Let me get back to the big bang for a second, because this splat theory not only brings out the requirements for multiple unseen universes, it also explains away the problems with the Big Bang.

- It allows for the slight ripples in the cosmic fabric that created the seeds for the formation of galaxies and large-scale structure in the universe, which goes against the Entropy model without destroying the Law of Entropy.

- It accounts for the absence of super-massive particles known as monopoles, which the intense heat of the early Big Bang should have produced in great amounts. I know no one has ever found one of these mysterious things, but Big Bang needed them and the old Ekpyrotic Universe delivered them.

- It insures that the necessary particles never had to reach infinite densities like the original Big Bang model required, but could not explain.

Of course, this splat theory doesn't solve all of the problems and makes a new one in trying to quantify this parallel, unseen, "universe".

Second Universe

I know you would just like to confine reality to this universe so you can sort of hold on to what you think is reality, but it won't make it go away and this unseen universe thing actually makes things fall into place better than if we ignore the probability. It also allows for a reasonable backdrop for vibrational matter so let's look at a second universe for a little. This second universe of the "big splat" is further defined by another new theoretical model called super-symmetry that will be addressed in more detail later. Super-symmetry essentially describes a connection between the two universes. According to this model, things did on one universe affect things done on the other. If too much change is forced on one universe, the inhabitants, yes inhabitants may get perturbed. I call the inhabitants angels [Assuming one calls the second universe Heaven], but you can call them whatever you like, because I am only providing the data. Certainly, matter is affected between these 2 universes just like everything else. To make it easy, let's just say that if matter appears in this universe, it must have been lost in an adjacent one.

Once we understand how the universe actually operates, we can more easily understand weird stuff like flying and levitation, disappearing matter and people, changing lead into gold, turning a stick into a snake or even changing a person's appearance into some type of reptilian featured person like the one that gave Eve some tasty fruit so long ago. All of these things have been reported repeatedly by people around the world and throughout time. While we are continuously told they don't exist and that anyone who tells you about these things is a nutcase, we are now finding out that we have done our children a major disservice telling them that many of these things are untrue.

John Keely's Vibrational Sympathy

In the 1880s John Keely came close to the truth. During his investigations, he invented numerous devices that seemed to channel vibrational energy. He claimed to have discovered a new motive power which was originally described *as "vaporic" or "etheric" force, and later as an unnamed force based on "vibratory sympathy", by which he produced "interatomic ether" from water and air. He refused to reveal the secrets of his inventions and methods. In 1884, his "Vaporic gun" was demonstrated. His description of what it did is very interesting to use. He stated, "I take water and air, two mediums of different specific gravity, and produce from them by generation an effect under vibrations that liberate from the air and water an inter atomic ether. The energy of this ether is boundless and can hardly be comprehended. The specific gravity of the ether is about four times lighter than that of hydrogen gas, the lightest gas so far discovered." [New York Times, 22 September 1884]* Certainly, he had some things wrong, but he was on the right track by investigating the vibrational component of structure.

OK! He mostly put a bunch of long words together, but it's the vibration and ether things that are of interest to us here.

He also indicated, *"It is an elaboration of inter-atomic ether by vibration. The **atomic ether vibrates all around the molecules of matter**. There is a magnetic force attached to it at the same time, and it assimilates with the molecular atomic aggregations - that is, assimilates with a certain attractive force that it is hard to tell what it is. I call it a **vibratory negative**. It doesn't act like a magnet drawing metals toward it. There is a certain magnetic effect about it that causes it to adhere by vibratory rotation to different forms of matter - that is the molecular, atomic, etheric, and ether-etheric. The impulse is given by metallic impulses, the rotary power that is formed by etheric vibration - that is the force that holds it in position."* [*New York Times*, 7 June 1885]

He was a nutcase and later in his life he was considered a fraud by some, but his views may have been the seeds needed by one of the greatest minds of recent times when it comes to vibrational matter and effect.

That guy was Nikoli Tesla. We'll get into his work in a little; I just wanted to give you a glimpse into how long these studies have been going on. I'll try not to get as wordy as Mr. Keely, because I simply cannot understand what he was talking about most of the time. Later, I'll readdress this pioneer. First let's look at new discoveries in sub-atomic particles. What we are actually finding is that the atom is not a good reference to determine materials. We're used to atoms so we use them and reference to them makes chemical equations easier to define, so we won't throw them away, we just need to look deeper.

Sub-Atomic Particle Research

If you have wondered about anomalous characteristics like levitation, disappearing and reappearing, and even spontaneous human combustion, they are all supported by new developments and discoveries accomplished by something called the "Tamashii Project" of subatomic particles and other similar projects and theories. Before we investigate details, let me say I'm sorry for the technical detail. It is not my intent on making equations fly around here so most of the data will be greatly generalized to make is simple. It is, however, important to open your mind to this concept. Without this backup, some of the details presented in this "liberation of concept" will not make sense and probably will be disregarded as too fanciful.

Sub-atomic particle research is a new science, which goes a long way in showing how levitation and most seemingly unexplainable phenomenon are possible. It also describes how photons miraculously appear, how atomic structure is determined and held, and how mass can be added and subtracted without time dilation. Everything we thought we knew has collapsed around us over the last few years and what has emerged is easier to understand because it doesn't have to have anomalies that could not be satisfied by the general knowledge".

Old Atomic Theory

In the past, we were taught atomic theory. In that basic theory, the atom was the smallest building block and all things are made from 115 different types of atoms. Scientists started to punch holes in the theory. They found Gluons, Bosons, Gravitons, Quarks, Photons and other particles much smaller than atoms. Here is the most fascinating part to me. Sometimes these particles just disappear. All the atomic theory was in jeopardy, but the theory continued to be taught.

Photons

Like I mentioned before, people started to look at the make-up of "light" and asked, "Where does a photon actually come from in atomic theory anyway?" What I was essentially told in college was that a photon was sometimes a particle and sometimes an electromagnetic wave. Just believe it and don't ask questions. The questions are only now starting to get answers and some of the answers make it look like levitation and even element conversion are both possible. This is neat! Maybe the ancient idea of changing lead into gold wasn't so wrong!

After all, the colleges are teaching that particles can miraculously convert themselves into electromagnetic waves. How much easier is it for a particle to simply change into another particle?

In school, you may have been taught that if a photon's vibrational period is 1×10^{-6} seconds it has special properties called "Infrared". Even if you didn't take that class, the photons are infrared. I just brought up college to make you think I was smart. Of course, infrared isn't red at all. It's invisible to our eyes. It should have been called Infra-invisible. It was named a long time ago so we are stuck with it. The different colors of light miraculously mix together and somehow turn into white light even though you always thought that mixing colors together should make the colors darker and darker until everything was black. While you see white light every day, there is no such thing as white light. It is simply the combination of a whole bunch of colors. As a photon's vibration increases to a period of 4 x10^{-7} seconds per cycles, it suddenly becomes "visible". As it speeds up even faster, the photon becomes an "X-ray" which can see through just about everything and then a "gamma ray" that can destroy tissue. The previous chart showed these things, but in this case I want to bring out something strange. All this instantaneous changing and complete modification occurred in a tiny particle and it was accomplished [**without putting in huge amounts of Fusion energy**]. I know it sounds absurd, but people accept it every day without questioning. Here is my question to you, "Just how does the photon particle change to another substance with entirely different properties?" We never answer this "simple" question. We just call it a photon. On the following table are the common "frequencies" of photons that we accept without question.

Description	Cyclic period	Freq. [Hz]	Characteristic
Helpful Infrared light	1 x10^{-6}	30 x 10^{13}	Invisible thing
Visible light	4 x10^{-7}	75 x 10^{13}	Visible thing
Dangerous X-rays	1 x10^{-8}	30 x 10^{15}	Invisible thing that penetrates bone
Deadly Gamma Rays	1 x10^{-9}	30 x 10^{16}	Invisible thing that destroys

You would think someone would stand up and say that is a lot of malarkey but almost no one does.

New research tells us that a Photon can better be defined as nothing more than a Boson emitted from a particle-group that has absorbed energy. Because of the energy boost, it must now eliminate the energy to insure stability-and here is the most important part. The most common energy absorbed is something we call gravity. You'll be shocked to find out that gravity vibrates just like everything else. The vibration level defines the gravity.

Therefore, the typical way to make light is to make an object appear to have less mass and, therefore, have less gravity. [I'm still way confused, so let's investigate some more.]

Don't get too bogged down in this initial stuff, because it will begin to make sense very soon. Besides the gravity connection here is an important part to consider from the training you got in school. If you look back at the first chart, "particles" vibrate really, fast and electromagnetic wave photons vibrate much, much slower. The block of light we are describing goes in and out of the particle-wave things all day long. I know you still don't know how the frequency changes, but at least you know that a photon CAN be sometimes a particle and sometimes a wave ----if the vibration changes very quickly.

Atomic Fusion

Here's an oddball question; if hydrogen has one electron and one proton and helium has 2 electrons and 2 protons, can you put 2 hydrogen atoms together and make helium? The answer has been, "you can't do it without the exchange of a substantial amount of energy associated with "Atomic Fusion". It has something to do with what we learned in school that was called NUCLEAR force [The force that "allowed" atoms to stay with large numbers of protons and electrons.]

According to the dictionary, a nuclear force (or nucleon-nucleon interaction or residual strong force) is the force between two or more nucleons. It is responsible for binding of protons and neutrons into atomic nuclei.

People create massive atom splitters to break the Nuclear force and make subatomic particles of all types. As the atoms are degenerated, there is a fear that the energy created during the separation of atomic particles could cause disaster or create a fusion disaster like a nuclear bomb or other scary things. . One of these cyclotrons in Europe generates so much energy that there were concerns that they might create something called a black hole. If you think I'm crazy, you had better stay away from the guys trying to explain just what a black hole is. If one of these black hole things happened, the earth could explode and we would all be sucked into the next universe and if any survived the trip, who knows what we would find. Let's not think about these black holes and try to stick with things that are more manageable like nuclear force.

Unfortunately, or fortunately, nuclear force can be controlled. The force that holds atoms in quantized numbers of particles with quantized amounts of energy and density apparently is a vibrational force. Here is one of my interests in this whole thing. Maybe we can get around the fusion reaction requirement that is so very dangerous to our existence with vibration.

Non-Fusion Manipulation

In a vibrationally controlled world, trying to modify this "APPARENT" "nuclear fusion" is not the only way to manipulate atoms. One can affect atoms by affecting the characteristic vibrations of the component particles or by affecting the apparent vibrational element of the entire atom {particle group}. These manipulations can and have been done without the huge explosion of a nuclear bomb. In fact, they may happen as part of a natural affect which emits photons.

By modifying particular frequency components of the atom, the various groups of sub-atomic particles apparently can and do become invisible, or will retrograde to another state, or may lose or gain their associated particle-mass energy by emission of that elusive photon thing.

That doesn't sound like light but it more closely defines how it is produced. I know that is a big statement so I'll get you more confused by defining invisibility. You can go back to just saying if this light "thing" vibrates faster its structure changes if you want. I don't care. The thing I want to bring up right now is that invisibility is somewhat different than you previously thought.

Invisibility Observation

Many times, invisibility has something to with another weird thing we call gravity.

All particles and groups of particles have this gravity thing, but not all particles have electromagnetism, so they may not all react to one another nor would they all produce magnetic fields. In fact, if two particles don't react to one another, they are invisible to one another. One way to think of this is that they are, in some way, shared between 2 universes.

Gravity Observation

Gravity of these particles may also be shared as a remnant of mass. One may believe that if gravity is sensed, by definition, mass must be where the gravity is. Some have tried to explain away gravitons, and other seemingly massless particles that have gravity, but without mass defined some way, the details fall apart.

You can't shield this gravity thing because the shield would have the same elemental particles. We are now discovering that there is another way to change elements besides changing the vibrational component.

A way to change elements is by modifying the gravity part of a particle just like I defined in the section on photons.

[Sounds simple; Right?]

While no one actually has defined Gravity in any major way, I think we can define it fairly simply [Ha!] in the vibrational dominated universe. Gravity would be the cross member to the vibrational string in a vibrational membrane. I guess you know what that means! Like everything else, gravity and vibration is the same thing. If we attach particles to a vibrational string, perpendicular vibration would show how the vibration travels perpendicular to the vibrational string OR [and this is a big or] ----

I'm sorry! I'm Sorry! That whole thing just blurted out of my fingers. I'll go slower and I'll make pictures so that I won't sound like that John Keely character I introduced earlier.

Cross Modulation Gravity Vibrations

Gravity can be considered the **vibration of a vibrational string**. We could recognize this gravity effect as a cross modulation of the dimensional string that makes particles as shown below. As we convert this structure back to the frequency domain and flatten it to a membrane. The modulated flat pattern arises. Later I'll show this in sort-of a space-time coordinate system to make sure you don't get confused.

41

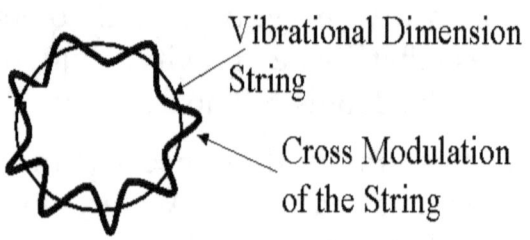

Vibrational Dimension
String

Cross Modulation
of the String

Graviton

A graviton is a particle that has gravity and no mass. Scientists see the effects all the time, but really defining what a graviton is can be a mystery. One way to look at the graviton thing is 2 fermions [Quasi-mass] vibrating at the same frequency and in opposite phase. Everyone knows what happens. It's like noise canceling in those BOSE headphones. In this case the particle becomes invisible even though there are 2 of them just sitting there.

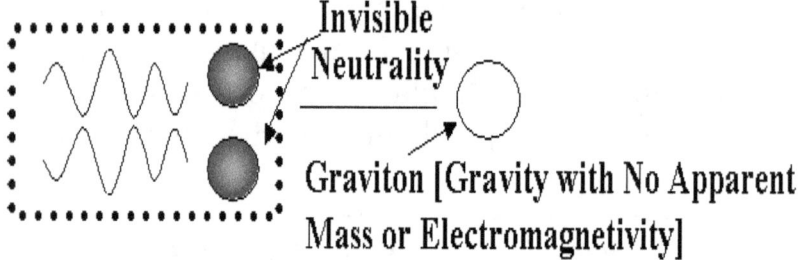

Invisible
Neutrality

Graviton [Gravity with No Apparent
Mass or Electromagnetivity]

Noise Cancellation

Let's take a nice set of BOSE noise canceling headphones. These things work by bringing in background sound, inverting the sound and pushing the normal sound and the inverted sound into your ear 180 degrees out of phase just like the drawing above is showing 2 "quasi-particles" vibrating 180 degrees out of phase to one another. While you would think there would be more sound for your ear to deal with, the background sound disappears just like the mass in a graviton.

Similarity To Electricity

In electro-magnetics, the electrical field component reacts in a perpendicular direction to the magnetic field component. This effect in space-time can be similar to the perpendicular gravity field that may be compared to something we should call mass.   Everything is vibrating. Gravity, Mass, Electricity, Magnetism, light and even the so-called nuclear force are all just variations of vibrational "stresses" to the elusive single particle that makes up everything and that brings us to Einstein again.

Einstein was perhaps the premiere relativistic thinker of the present day. While most of his discoveries dealt with functional groups of particles that made up elements and how these elements reacted to each other either in this timeframe or as it was modified as the dimension of time or its reference was modified. Einstein also studied squishy things.

Einstein

One of the first people to know there were problems associated with the "Clean" atomic theory was Einstein. He knew that if the electron cloud around a Protonic nucleus was to be held in place, there must be an unseen group of things between the nucleus and electron particles. John Keely had called this an "ethereal force" so Einstein decided to come up with a new word.

Einstein's Squishy Things

He called this collection of things the "Aether" and he said that it was made up of invisible, sort of squishy, particles that bounced around between other charged particle systems such as the electron and proton. These "Aether" [POTENTAIL] particles are invisible and have no mass, or at least none that we can sense. That was only the beginning. More and more data was collected which allowed a better understanding of the dynamics of the atom and even the electron itself. Einstein's Aether was actually starting to make sense, but Einstein didn't stop there. He couldn't.

After all, he had now completely shattered his $E=MC^2$ formula with these massless particles.

If you remember the energy equations from before, all were of the form $E = 1/2\ AB^2$ so his universal law of mass-energy should be re-written.

- E= ½ MC ^{2} [possible universal law of mass energy that goes against Einstein's original one.]

Does it make sense that the mass energy model would be different that all the other energy models? No! Everything seemed to fit in the old $E=MC^2$ until Einstein's invisible squishy things came along. Let's assume that the added ½ in the equation must be accounted for. What we find in Einstein's squishy things concept is the rest of the elusive picture with all matter and apparent affect. They are, simply, augmentations of a unified particle. [Later I will get into the 10-dimensional universe and may have to eat my words about Einstein.]

Unified Particle Theory

Einstein didn't change his original mass energy equation, because it seemed to work for most things but he did come to the conclusion that there must be an error, which allowed for one unifying particle rather than the hundreds that were continuously being found inside the atom. This unified particle theory is the basis for the new atomic science.

The NEW SCIENCE is looking into the details of the elemental parts of atoms. The current studies are getting close to finding that single particle that makes everything. When I say everything, I mean everything including matter, light, and even gravity. Let's start by looking at some of the tiniest known particles. Again, I apologize for the details, but it will come together in just a little bit.

The Huge Electron

An electron has a mass and movement; therefore, it has a gravity and electro-negativity. Scientists call the combination of these features "Electro-magnetivity". The electron can be further diminished into particles called "quarks" and "positrons", but as we go farther down into the individual elemental parts of the electron, the basic component, which we now called the "Boson" contains multiple fermions [quasi-masses missing components that allow them to be visible.] and **a frequency**. The reason for saying quasi-mass is that fermions making up bosons are invisible and Bosons, made up of invisible fermions, **can become invisible** just like a photon is before it is excited. Similarly, we can say it is like a graviton is when it is sensed and like neutrinos when they are generated. By placing a number of these "bosons" together, it becomes a "quark". This quark thing then produces either, or both of the forces of the universe, gravity and/or electro-negativity. Put enough quarks together and you will get the huge electron which can change electro-negativity into electro-magnetivity. Let me review this string of events.

- *Aether Vibrates to become Fermion [A variety of sub-particles are born]*

- *Fermion connected to the BOSON [A vibrational Particle is born.]*

- *Boson connected to the Quark [Electro-negativity-electricity is sensed.]*

- *Quark connected to the Electron [Electro-magnetivity-magnetism is sensed as outside components of our universe apply stresses to the Aether.]*

How to force the Bosons together is the secret and that is where the Tamashii model helps us. The model and the discussions that use the model as a beginning have changed the concept of reality.

The Tamashii Model

Now that you have a basic understanding of the Boson, I'm going to quickly go through the general Tamashii model. This model expresses all elemental parts of atoms, including the quarks and bosons, as combined groups of particles which change characteristics depending on their motion. My version certainly is much less detailed than the complete model introduced to the world by J. Newcombe Hodges some 10 years ago, but I think it will be enough data to bring in my own point of view. Even if you don't understand the BOSON yet, don't worry. It's one of those things that sort of grow on you the more times you say it. I'll say boson from time to time in this book so you will get comfortable and happy.

Tamashii Invisibility

As I stated above, correct quantity and frequency of Bosons makes things invisible and when I say invisible I mean containing absolutely no mass. I know you have been told every physical thing has mass, but it is not correct. Light, for instance, has no mass and neither do particles called gravitons. If particles in close proximity have similar motion {frequency} they can react with one another. Particles with different frequencies are unseen by adjacent particles and do not necessarily react. This may also mean that the particle is completely invisible to an outside observer. Because of this elemental property even a boson-sized particle can go all the way through the earth. The earth would not register any mass with respect to it and the boson would register no mass with respect to the earth. The mass of each simply would not exist to the other. Stay with me here because that is sort of what causes light as defined by these new sciences.

Just think of mass an illusion of vibration and you'll be fine.

Tamashii Light

Everyone knows that if you put two frequencies together, the combination is more than the two frequencies. It also contains "beat" frequencies that are associated but different that original two. According to the Tamashii Model, collisions of these two "different frequency" particles do the exact same thing. Interactions can cause secondary "Beat" frequencies to be generated. This "beat" frequency is the photon emission [light]. So, in general, the close proximity of two different frequency particles is what causes light.

Just think of light as an illusion of vibration. Remember that you can't really see light. All that the brain gets fed is the vibrational content of these photons. The brain changes the frequency of the photon vibrations into colors and the abrupt change of color allows us to distinguish things.

This is a little easier to believe than- "sometimes light is a massless electromagnetic wave and sometimes it isn't because it has mass and is a tiny particle." While this is the "Normal" way to describe light, it really is a stupid way. That brings us to water.

Water

We used to think that water was made up of 2 parts hydrogen

to 1 part oxygen, but many studies since 1995 have now concluded that the ratio may be better approximated with a 1.5 to one ratio during some time periods and the full 2 to 1 ratio at other times. The first realization came at the ISIS neutron spallation facility in the UK where German-British collaboration collided epithermal neutrons. By determining the energy loss after collision with water molecules, it could be concluded that for short time periods [on the order of 100 attoseconds (1 attosecond=10^{-18} seconds) that there were 25% fewer Protons in the hydrogen ion mass.

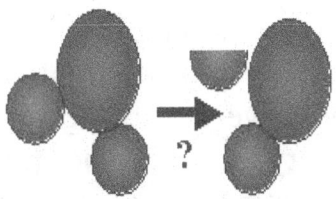

Additional experiments with carbon compounds $C_8H_{14}O_2$ and C_6H_6 over the years have shown a similar anomaly. At Uppsala University in Sweden and the Australian National University in Canberra, the experiments were changed to detect electron counts rather than proton counts and the results were identical. The results of all of these experiments can't be possible if we use "normal" models, but if the various elements could disappear from time to time, the conclusions become more reasonable. The Tamashii model again can be used to justify the findings.

Quarks

Changing the frequency can change the characteristics and union possibilities. What this means is that the huge particles we previously thought of as building blocks of matter, atoms, are made up of still smaller particles that can change because of some outside influences. This "change" can and often does include invisibility.

As we get inside the atom, if 3 or more of these tiny, tiny particles, called Bosons, join, they become one of 6 different types of Quarks [Charmed, Strange, Up, Down, Beauty, or Truth]. These are weird names for weird particles. Odd quantities of Boson particles are visible and appear to have mass; even numbers have no apparent mass because they are electrically neutral and don't tend to interact outside themselves.

Gravitons

One such particle is the graviton, which has no apparent mass, but has a gravity, which requires mass. [Ahah!! We are talking about things that are so small that we can barely imagine them and now we are trying to define the universe with them.] This "theory" ties together photons, wave theory, particle theory, gravitation without mass, and these new discoveries of molecules changing characteristics. We at least have a chance at having a real definition of particles with the Tamashii Model rather than the usual "definitions without details" we tend to accept. When I say "usual", I mean when a text book simply states "therefore it can be concluded" and you read before and after the statement and there is no way that the result can be concluded except by defining away many exclusions and exceptions.

A photon is NOT sometimes a particle and sometimes a wave like you were told. Both a graviton and photon are made up of the same particle. The quantity of particles and their frequency make them invisible or visible depending on what they come in contact with.

Let's say there are 2 fermions[quasi-masses] minding their own business and they happen to be vibrating at different frequencies. What happens is that there is a beat frequency produced and the particles become visible in a string with that vibrational pattern, but there is something left over. If 2 frequencies beat, there is a negative and a positive beat frequency [known as a positive and negative beat frequency one becomes visible as a particle while the other generates electro-magnetism or a PHOTON. The diagram following shows how this function might be interpreted in a "time-space" way.

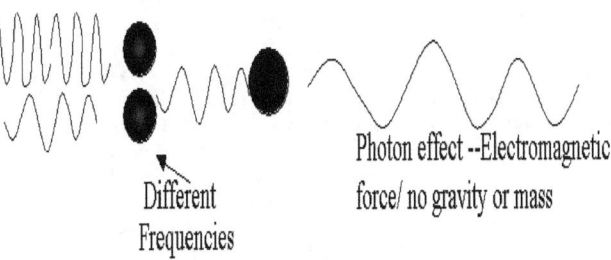

Different
Frequencies

Photon effect --Electromagnetic
force/ no gravity or mass

Particles

All we have to do, according to this model, is look at the different particles and we will know everything, but they are really small. Some turn into light, some exert gravity, and some have no mass. Presently these particles are called Bosons. Three bosons combine, they become quarks and depending on how they combine, the quarks act differently. Scientists have classified many of the quarks and given them very impressive names. Multiple quarks make other characteristic particles on-and-on until the atom is finally established. I came up with a short list of "defined" particles so that you would not get any more confused than you already are. Following is the short list.

Fermionic Particles

Fermion	A particle that has apparent mass and exerts gravity and electromagnetism. [Electrons and Protons could be considered fermions; however, many today use the fermion name as a particle missing a component needed for it to really exist like a graviton. It has gravity, and must have mass, but no one can find its mass.]
Quark	This is a class of Fermion with something called ½ spin characteristics. There are six types named Beauty, Charmed, Strange, Truth, Up, and Down. Each is made from either three leptons or three neutrinos.
Neutrino	This is a quasi- Lepton that is a component of an Up-quark, three types known [Electron-neutrino, Muon-neutrino, and Tau-neutrino.] **They have almost not reaction with matter and can pass through the Earth**--- They have no apparent mass.
Photon	This is sometimes considered a Boson that has no apparent mass so it really is a fermion, but has electromagnetic properties. It is made up of 2 quarks or equivalent particles. It exhibits no Gravitational force, but instead it makes light.
Graviton	Like the photon, this is another fermion possibility that has no apparent mass, yet it exhibits a strong gravitational force. This suggests that an even quantity of quarks is combined in its makeup.

Bosonic Particles

Boson	There are several different kinds of this particle. Some affect gravity, some effect magnetism, some effect nuclear forces. One theory is that these occur when matter and antimatter match and come in contact with one another. They sometimes exhibit no apparent mass but still produce a magnetic, gravity, or other force. The Photon and Graviton are sometimes considered bosons.
Gluon	This is a **boson** and has no mass. It is a particle that can cause an interaction between Mesons. It exhibits a strong nuclear force. These are passed back and forth inside an atom's nucleus. There are 8 different types of these particles known.
Lepton	**Bosons** that do not attract nuclear force. The electron is the simplest form. There are 12 known varieties of this particle with 6 of them having no mass. [Yes Leptons without mass may be considered fermion.]
Tau	This is a heavy Lepton that is a component of a Beauty-quark.
Muon	This is a heavy Lepton that is a component part of a Strange-quark; this one is found in Cosmic rays so stay away from Muons.

52

Baryonic Particle

Baryon	This is a particle with 3 quarks, so it's usually heavy. The only stable Baryons are protons and neutrons. They have strong nuclear force. There are at least 10 unstable types besides the proton and neutron.
Hadron	Quarks, mesons, and baryons that attract nuclear force end up being called hadrons.
Electron	A huge quantity of particles [3 full quarks] in a mass. This is the smallest stable particle.
Proton	This particle is made up of two up-quarks, and one down-quark, because an up-quark is much more positively charged than the negative charge associated with the old down-quark, the proton looks positive. This is a Baryon particle.
Neutron	This particle is made up of two down-quarks and one up-quark. Because the down quark is half the charge strength of the Up-quark, the particle is electrically neutral. This is a Baryon particle.

Anti-Matter Particle

Positron	**Antimatter** equivalent to an electron.
Anti-neutrino	The antimatter cousin of a neutrino, this particle is **VERY important** as it makes Nuclear Force. As a neutron decays to an electron and proton, the antineutrino is what allows the release of energy in the form of momentum while still allowing for conservation of momentum to be satisfied. Without it nuclear power cannot be explained.
Meson	A particle made from a quark and an anti-quark is a meson. Because the various quarks have different electrical charges, these particles can also be electrically charged. There are at least 13 known varieties of this weird particle.
Pion	This is a combination of particles and anti-particles which together cancel out magnetic and gravitational properties.
Kaons or K-Meson	These particles are the transition of a Meson into something else. Once one of these babies is made, the antimatter sister become a neutral particle when coming in contact with a normal Kaon, but the reverse does not occur. Nobody knows why.

Antimatter

I guess you noticed I slipped in a new thing. You can see from the list above, that some of these newly defined "things" are made with anti-matter, therefore, the simple definition of antimatter should be known. The latest definition is the following:

Antimatter is any particle moving backwards in time.

For me this is difficult to imagine, and I'm not going to get into the time reversal thing in the first part of is book too much, but you must realize that an adjoining universe could certainly experience time in a backward direction to the time we experience. Even without backward time we are not done yet, so please hang on just a little bit longer. According to this new science, normal matter particles encountering anti-matter particles become those little rascals called bosons, and you possibly guessed it, the particles disappear, but they are still in existence because they still have electro-magnetism and gravity.

Besides this disappearing act, there is another strange thing about this anti-matter stuff. We have not been able to locate very much of it and you would think that there would be the same amount of each type particle at any given time period, sort of a conservation of particle similarity law or something. That's where the "M" theory takes over.

Antimatter and the "M"

By all reckoning, the antimatter stuff is mostly located in an adjacent universe according to the Big Bang theory and its corollary called the Membrane Theory, which is almost always referred to as the M-Theory. You would think they would give it more than just a letter but they didn't so we will have to live with it. I need to give you a glimpse of this M-theory for a number of reasons. Not only does it allow for the existence of the other universes including a place called heaven, it also explains some of the basic components of weird things like Alchemy. Even more importantly, it shows that ideas like alchemy, invisibility, transmutation of living matter, and levitation are not as bizarre as the more "normal" sciences. In fact, it shows us that these things may be more common that one would normally believe. The levitation thing may be better explained by looking at the weights of particles.

Wrong Weights-

To make matters confusing, scientists have weighed some of the particles by testing interactions with something called the Higgs Boson. Following is a short list of some of the more common particles showing comparative sizes and charges. I think you can see from the data that you cannot recognize a particle by its mass, spin or charge. The best way to levitate something is to make its mass "appear" to be zero. The opposite also is true if you want something to be heavy make the mass "Appear" to be high.

Particle Name	Charge	Spin	Mass [GeV]
Photon [Fermion]	0	1	0
Graviton [Fermion]	0	1	0
Gluon [Fermion]	0	1	0
Z [Boson]	0	1	9×10^1
Higgs [Boson]	0	0	2.5×10^2
Muon Neutrino	0	½	2.7×10^{-4}
Electron Neutrino	0	½	7×10^{-9}
Tau Neutrino	0	½	3×10^{-2}
W [Boson]	+/-1	1	8×10^1
Electron	-1	½	5×10^2
Muon	-1	½	1×10^{-1}
Tau Lepton	-1	½	1.8×10^0
Up Quark	2/3	½	5×10^{-3}
Down Quark	-1/3	½	9×10^{-3}
Strange Quark	-1/3	½	1.8×10^{-1}
Charm Quark	2/3	½	1.4×10^0
Bottom Quark	-1/3	½	4.5×10^0
Top Quark	**2/3**	**½**	**1.8×10^2**

One of the more confusing particles is the Top Quark. Look how much heavier it is than the other quarks even though it is essentially identical to the others. Many researchers ignore the anomaly, but if you can arbitrarily change mass, you can arbitrarily change weight and levitation happens. Without the addition of "frequency" as the controlling component of charge, size, weight, light, magnetism, gravity, and electro-negativity there has not been a good answer to this dilemma. Even with the use of vibrations to change masses and particle characteristics, we still have a problem changing masses unless there is something else going on.

If mass changes in this universe and mass is a constant, where does the mass go?????

That's where the super symmetry theory allows us to explore the possible answer and at the same time allow for a critical connection between our universe and the one we typically call heaven.

Super-Symmetry Answer

If you remember my introduction to the M-theory, suggested that 2 universes were required to form our particular universe and its existence was required to make the whole big-bang thing work. Well! There is more to the dual dimension M-theory than I previously detailed. This concept may help us with the anomalies of particles being found. Part of the answer has recently been formulated by a system called super-symmetry. This model can be used in conjunction with the Tamashii model to give us a way to modify masses and charges. In this model, large particles have a sister particle in a secondary, unseen, parallel universe. If I didn't make you lose interest by bringing up multiple universes again; here is the meat of this model.

Everything is symmetric by constant mass.

That is, a large particle has a sister particle that is tiny while a tiny particle has a very large particle. Examples are given below:

- Photons [large bosons] are partnered with Photinos-tiny Fermions]

- Quarks [tiny Fermions] are partnered with Squarks [large bosons]

So long as the particle twins have the same mass as other particle twins, all is fine in both universes. Everything is symmetrical. If a particle is changed in size due to some frequency agitation, the corresponding "other universe" particle simply changes in the opposite way. I know that all this parallel universe stuff is confusing, but scientists today are having a more difficult time **not** believing this feature of nature. Make no mistake; this parallel universe is totally different than what is here and actions in the "other place" are not governed by what is done over here except when basic particles are changed from one substance to another. If people in this universe started changing things as they pleased, the beings living in the other universe might get pretty upset. By all accounts, in the olden days, this curiosity and many others were well known.

Particles and Life

It is expected that changing life would have the same super-symmetric relation and changing of particles. I will briefly review that concept, but generally, life and the meaning of life must be dealt with separately. It was as if they were separate dimensions.

Antimatter & the M-Theory

If you are wondering where all of the antimatter is, the answer by super-symmetry is that the other universe has much more antimatter particles than what we believe to be "normal" particles so there is no need or reason for this universe to have more antimatter. Modern, mainline physicists have proven this mathematically with the Super-string theory, which is a subset of the M-Theory. Actually, the M-theory revolves around 10 [or more] dimensions and

58

vibrational nuances that change the size and entropy factor for closed loop strings. Larger "circles" would have less energy and have a lower vibrational component and the smaller the circle of a closed loop dimension. Some try to identify open loop dimensional strings, but that gets me way too confused so I stick with circles. Another way to state this is that a larger "circle" would be time dilated over a smaller circle. That stuff doesn't matter right now, but I did at least want you to see that "normal" scientific modeling requires the use of frequency modeling for the definition of matter and the definition of a "dimension"

Stringy Details

Let me try to identify some of the elements of the M-theory by first going over some principle of normal Sting Theory if there is such a thing. String theories typically are defined in time-space rather than in the frequency domain, but in string theories, we don't just have height, width, length, and time that you are used to. Universes actually breaks down the cosmos into 12 dimensions interpreted as ten different one-dimensional strings. The strings can loop, spiral or simply wiggle around.

Whenever two of the strings come in contact with one another, they may or may not interact depending on how they are vibrating. The following graphic, sort of, shows how they may react to produce 4 dimensional planes or membranes. The reason I am showing these things is not to confuse you, but to give you the idea that these things flex and twist and turn and determination of reaction is controlled by the perturbations. The perturbations, like just about everything else is dictated by some vibrational component of the dimensions making up dimensional membranes. The more the perturbations are enhanced, the more the energy level is sensed. The larger the particle becomes, the more work that is done by a system. While all that seems OK. Make no mistake; vibration does it all.

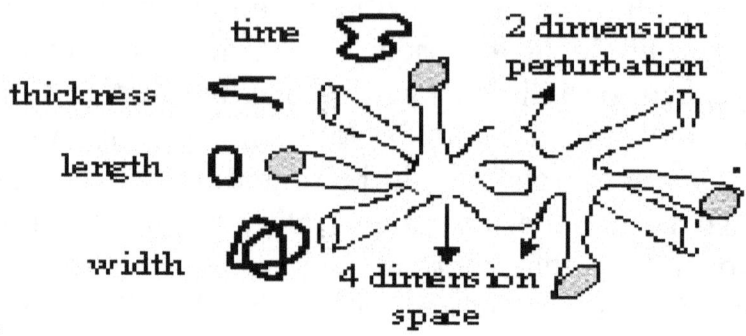

Now just imagine what 12 dimensions would be. According to most current science theories, only four are used for our visible universe unless you add the other 6 that are sitting around in our universe doing nothing. Scientists are saying the others are all in our universe, but they are compactified so we can't see or use them. I know that simply doesn't sound right, so we will get to that absurdity in the later chapters of the book. Right now, let me get back to confusing you a little. I'm sorry for the confusing, but potentially within your confusion, a little more truth will be revealed to you. I will try o keep it as least confusing as possible if you promise not to put the book down. So here is what happens if mass is changed. Changing one "dimension" [changing mass, for instance] can destabilize another set of particles that are connected by similarity and produce a perturbation of dimension.

What that means is, the guys in the other universe would not like us messing with their reality.

Time could be affected or elemental sizes, or just about anything, we hold as constants. Wars would be fought, and the changes caused by the perturbation could become permanent or momentary. That is not an excuse for war, but it comes as close to a valid reason as you could come.

By this logic past wars between heaven and our own universe made sense so hang in there.

More String Theory

I know nothing is making sense right now so let's look a little closer into this string theory thing. I know it's the last thing you want to do right now, but I think it will begin to get clearer as we go along.

Without the String Theory math was turning evil. Atoms couldn't be atoms, particles weren't particles, matter wasn't matter, and the Big Bang theory didn't work.

All the scientists just looked for someone to blame. String Theory came along to get us out of the doldrums. This was to be the theory about everything. Without getting into the meat let me, at least try to bring out some new elements that should be considered when trying to eliminate anomalous events that have occurred in our past.

String Universes

First let me expand on the universes that are made of these one- dimensional strings. While that sounds like an easy thing to understand, what they are talking about are wiggling things that have a dimension like time, but nothing attached to them. While it sounds like the String Scientists are completely nuts, the theory, mathematically allows for just about all the anomalies encountered when we assume that all the dimensions had to be attached to one another all the time and there was only one universe in a block of space. Get enough of these impossible to understand strings interacting and you have a universe. Because the reactions react to something, you HAVE to have multiple universes. The nutty part seems to be that all the universes have 12 dimensions with potentially 4 that interact and the other 6 that are simply not used and all the universes that are commingled interact with one another. [While this is now the common thought, in later sections I will demonstrate how all 12 dimensions are visible and useful in our universe.] Well it was just too much for a single theory to handle so corollaries started popping up that were distinct theories in their own right. We need to look closer at the corollaries and I know it's confusing. Hopefully, I will get you out of the confusion 'somewhat" as we go along.

More Super-Strings

Let me get back to the Super-string theories. These theories work best in those ten space-time dimensions I mentioned before. Because of the 10 to 12-dimension requirement, we only observe four space-time dimensions in our universe. The mathematicians got together and destroyed a good word by claiming that the 6 unseen or felt dimensions, are [here it comes] "compactified" so you can't get any reference about what they do so they aren't observable.

Take that compactification as a challenge. It doesn't make sense to even have the components. This introduces a new concept "observable physics". Because each universe actually uses all 12 dimensions, the 12 dimensions represent many different universes that are shifted from ours in one way or another so that we can't see of feel them or ANYONE in the alternate universe.

Are you hearing the "Twilight Zone Theme" in your head? While generally we can't see the other universe, the basic premise of matter has an issue. We know that matter cannot be created and yet matter seems to disappear and appear all the time. The super-symmetry thing indicates that for matter

62

to become "visible" in this universe, it must loose its visibility in another. Everything would be equal and opposite. I know that is sounding like it makes sense and it is scaring you that something might make sense here. There are actually 5 different super-string theories and I am certainly not going to try to mash the differences into your head because they only look at those string dimensions that seem to be impossible to understand in a different light. In these theoretical differences, some strings make circles, or are closed or open ones come around, or multiples tie together, etc. like I showed before. As far as I can tell, all strings should be considered as "closed" strings because of something called regeneration. Open strings would necessarily have a beginning and an ending. Not just the ends of the impossibly to understand string, but the actual extinction of the thing. What the strings look like is variable, but there is something that that we can hold our hats on. It is a condition of all strings and how they interact with those associated with them. This condition is called duality.

Duality

Super-string theories ALL use something called dualities and dualities help us understand vibrational matter, so I had better sort of define this word so that it won't be confusing. Putting it simply it means that you can look at the same phenomena in two different ways and the two ways are inversely related. Let me give you a quick example. If we look at a space ship going through space and we speed up the dimension we call "time", the "distance" traveled seems "and is" less. If we increase "distance", "time" per "segment of that distance" appears longer or gets slower. It is only a matter of perspective or viewpoint. This same duality easily moves into vibrational matter. The example I gave was confined to our own universe, but it seems that the same

63

thing holds true for interactions viewed by separate universes as well. To identify vibrational matter and its duality, we must first leave the Cartesian measuring method and go to a frequency measurement system of the entire universe.

Frequency Domain

For this detail let's take our 4-dimensions Length, Distance, Width, Height, and Time. From this an area membrane can be constructed. An area membrane [or dual dimension component] could be width and distance, or height and distance, or height and width. Later I will expand this to the active 10-dinensions that can be recognized as our truer universe, but right now, let's stick with the concept that you are comfortable with.

Vibrating Distance

To get the hang of string theory, think of a guitar string that's been tuned by stretching it properly to fit. Think of this as a distance dimension in a closed or looped String. Depending on how the string is plucked a different tone can be made. One mode of vibration could make the string [or group of strings] appear as a boson [real particle], another as a fermion [a quasi-particle missing a component needed to be "real" in this universe], and so on because it may react or not react to other systems on the string.

In string theory, as in guitar playing, the string has to be under tension in order to become excited. A big difference is that the strings in string theory aren't tied down to anything but instead are floating in space-time [Right now I calling it this because it means something to you. Later we will have to change the name of time-space.]. While the vibrations of a "plucked" string changes the "cross-sectional" length of the string perpendicular to the guitar, the shortest cross-sectional length is when the string is relaxed and it gets longer as the vibration causes the string to pull away from the guitar. As part of a duality, to the guitar, the vibrating string always stays the same length.

In this example, the guitar string can be considered as the universal particle, the amount of vibrational distance and frequency changes the effect of the particle [It becomes a boson, or photon, of electron, or gold]. Now let me put the example in 2-dimensional space as a membrane in a membrane. Hopefully this will make the concept clearer and describe how super-symmetry works in a slightly different way.

The harder the string is plucked the more energy is transferred into the string which can be sensed as more motion by the string. [More amplitude of vibration]

This vibration is only in the direction of the plucking as shown below. It has only one vibrational dimension.

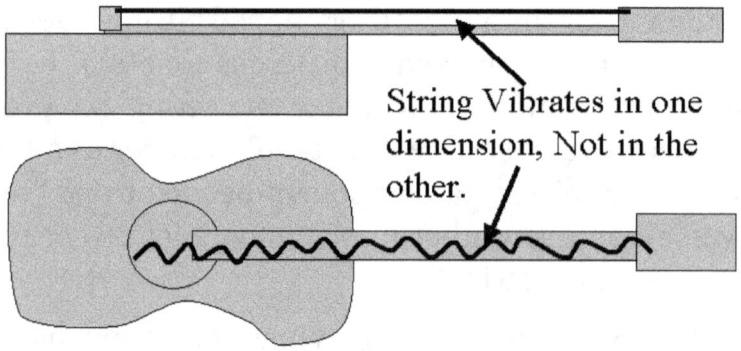

String Vibrates in one dimension, Not in the other.

Modulated "Area"

Now let's expand this just a little and bring in multiple universes by showing this modulation as an area. When trying to determine the AREA dimensional effect, the membrane made up of distance and width can be modulated and no size change would be noted. As an example, take a square piece of rubber and hold it at the 4 corners. Modulate the rubber by hitting the center [sort of like plucking the string]. It vibrates back and forth, but to the people holding the piece and looking down at the sheet of rubber, the size is constant. This back and forth motion is a duality. What I mean is that if the rubber moves in one direction so many inches on one side of the rubber, it moves the exact opposite amount on the other side of the rubber. Think of one universe being on one side and another universe on the other side. [This is the essence of super-symmetry in universe duality.] While both universes are affected, the actual universal elements are not affected in any way and the two universes are affected in an exact opposite way.

The duality difference can be defined by time moving forward and backward or the energy input being out of phase between the 2 universes [as one particle gets large, the symmetric particle gets small.

For this detail, let's take our 4 dimensions Length/distance, Width, Height, and time. From this an area membrane can be constructed. An Area membrane [or dual dimension component] could be width and distance, or height and distance, or height and width.

Closed Strings And Time Dilation

The above 2 examples could be considered as closed dimensional strings. In the guitar example, if the guitar that holds the string is rigid and if the two-people holding the rubber in the membrane example are actually one clamped fixture that cannot change its shape, the examples are of a closed loop so let me get a little weirder here and tell you about an issue with close strings. Because a closed string actually connects its 2 ends together to make sort of an "O" shape, vibrations are constricted in that dimension. In this dimension, the vibrations must have an integral number of wavelengths around the string. This means that as particles vibrate along a closed string, the vibrations must be such that when the particle reaches the location it started from, its position in its vibrational tract must be identical to the one it had when it started or it would not be able to exist in that dimension. If the components that make up the frequency and distance don't match the closed environment, the distance environment must be changed to ensure that the vibrations DO match. Again, we have a time dilation effect defined by vibration.

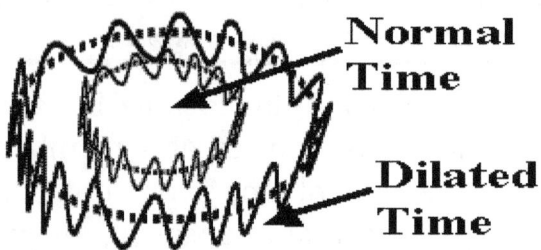

What you see on the preceding page is that in a closed universal "string", a particle exists if it has a quantum period of a whole integer as the number of cycles needed to encircle the host string. If the number of cycles in the string system are not changed but the apparent universal string is increased in size, those in an adjacent system see the even as a time dilation. While those in systems containing the dilated time element fell no difference, this type of dilation can certainly happen to any uncontrolled set of strings. That brings us to more membranes. The first diagram below shows how the example dimensional membrane looks when looking down on it. However, if we look at a cross-section as shown in the second figure, it can be seen that the membrane is vibrating in both directions. We can understand this to be a closed string if an integer number of cycles are resident along each edge. For this example, one direction contains only a single cycle while the transverse direction is made up of 3 complete cycles. For another component to fit within the confines of the membrane, it must have a whole number of vibrational cycles. One could interpret this as a **quantum element of this particular membrane**. The last diagram is of an abnormal event that may or may not be visible to one or both universes but it affects each in an opposite way. It is a duality associated with what we could term as a time dilation. One universe would experience dilation while the other would experience and compaction of time.

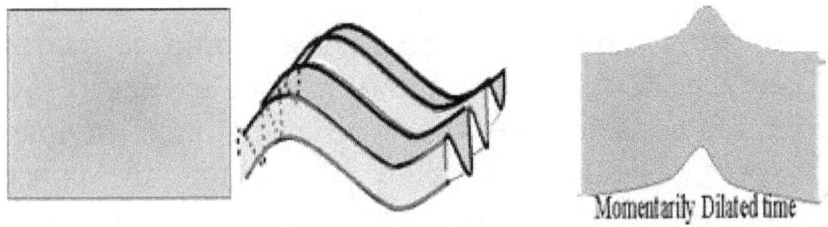

Momentarily Dilated time

Let me get back to the vibrating rubber example. Put a person on the rubber membrane that is vibrating. Now pluck the rubber [as shown in the diagram on the right above. When the vibration is at it farthest point from the relaxed

69

condition, time for him might be "faster'. He truly has a shorter distance to travel across the piece of rubber. As the rubber extends outward in its vibration, the distance is longer and time would be slower. I know I'm talking about vibration of a vibration, but that is exactly how our world works in a macrocosmic view and it is how the strings regulate things in a closed string pattern. One of the dimensional elements is "modified" to allow the loop to exist. In this example, "Time" is changed. When it is synchronized, no abnormality is sensed. While the dimension is closed, the characteristics of things can change. The plucking continues to be anomalous until the variations are synched up with other components of the system. The vibration of the vibration causes beat frequencies and those beat frequencies govern what is visible and what is not visible to our universe. The vibrations cause apparent creation of matter and elimination of matter or invisibility. Of course, the exact opposite would happen in the adjacent universe and that is where the Universal Existence Law comes in.

Universal Existence Law

If you disappear in the universe--- You will appear in another.

I know you have been told that when you die in this body, you will go to one of 2 places. Those would be: heaven or hell, and I'm not commenting on that one except to say that if your life-consciousness does disappear, it must reappear somewhere else. I know some believe that the conscious reappears in an animal, but let's assume that people don't turn into bugs. That just freaks me out. This overview is a macrocosmic interpretation, but a microcosmic one may bring out the essence of vibrational matter much better. Let's talk about particles and waves with respect to quantum mechanics on a micro scale.

Quantum Mechanics and Vibration

One of the most common dualities in physics caused the invention of "Quantum Mechanics". This fundamental consideration states that things can be considered as waves or particles. The more you describe something as a particle, the less you may describe it as a wave. What we are going to do in this book is to explain why this quantum mechanics thing works. The answer is that it is not simply a different perspective of the same thing, the wave or "Vibration" of a particle defines the particle. Increasing the vibration or decreasing the wave period as a quantum would be defined. Changing a particles vibration is actually changing the particle into something entirely different than it previously was. When a photon, for instance, changes its characteristics from a particle having mass to a wave having no mass, there is ACTUALLY a change in its characteristic, not simply a difference in viewpoint that is identified in Normal Quantum Mechanics. Particles can, at will, change how we perceive them by changing the vibrational component. In fact, what we are finding today is that our universe should, more precisely, be defined as something entirely different than Length/Distance, width, height, and time as the 4-dimensional blocks.

Vibrational Duality

As we are finding our how very much vibration controls everything, we also can determine today that vibrational components SHOULD be the defining elements of our universe and anomalous details begin to go away. While it is not an exact science here and no one can really see vibration, one could set up temporary definitions of a vibrational universe as described below:

- **Vibration frequency** [which provides the basis of time. It is how fast things vibrate]

- **Particle density** [the amount of mass used to define the universe itself]

- **Vibration variation** [Which changes the particle visibility, gravity, mass, attraction to other particles]

- **Vibrational travel** [the characterization that determine apparent size]

Time-can be considered as simply an average of the different vibrations in an area. If the vibrational baseline increases, the time dilates. If the vibrational <u>average</u> decreases, time constricts. Time is considered constant so long as the components are not excited in an abnormal way. If we artificially restrict the vibration differences, time will appear to slow down. If we need to stay synchronized in a closed dimensional string, time dilation will occur.

Distance – would be a combination of several of these dimensions as well- <u>particle density</u> to allow a reference of distance, vibrational <u>variation</u> giving the reference to our concept of time.

Width and Height/Area- would be the interaction of vibrational <u>travel</u> and <u>frequency</u>. As the phases align, the apparent lateral space taken increases. Given various phase relationships of a vibrating particle, all planer characterizations can be defined.

I know this is a lot to chew on. I've now destroyed the very foundations of your concept of the universe and I apologize from the bottom of my heart. By the way this really has little to do with the string things except that stings of the vibrational dimensions can be used just like the NORMAL ones to define our world and the interactive universes that are somehow integrated with ours.

As a brief concept, let's take an adjacent Universe. I know adjacent doesn't mean anything in my theory above, but let's just say a universe that uses the same dimensions, but visibility is depicted as how much out-of-phase vibrational components are. The graphic following shows "adjacent universes sharing a common dimension frequency membrane.

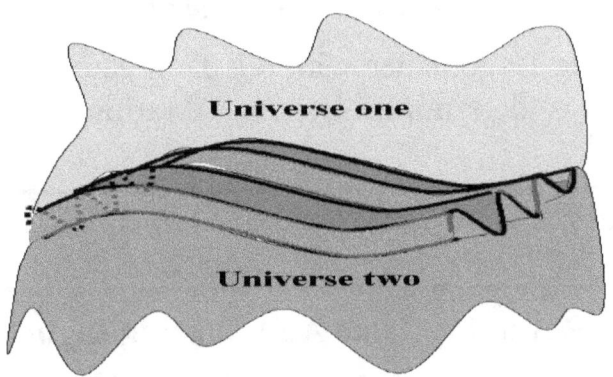

180-Degrees Out of Phase

As shown, all the frequencies of the 2 universes along the membrane are exactly 180-degree out of phase with one another. This out of phase thing is important.

If particles vibrate output phase with one another, the matter created by the vibrations doesn't exist to the other.

Another thing to recognize is below.

If one wanted to travel from one universe to an adjacent one, the only thing that is required is to invert the vibrational component with respect to motion.

Let's get into a little more detail about vibration components.

As particles vibrations get more and more in-phase, they become more visible in this universe.

I know you think you would see things slowly appear as they transitioned from one universe to another, but the "<u>closed</u> membrane characteristic" doesn't allow it. Because the "off" frequencies would not fit into closed looped vibrational membranes, a particle changing its phase would quickly sync up to the "new" phase and "new world".

Vibration and Energy

Another thing to consider with all of the frequency stuff is that energy is determined by the vibration in our various models.

The slower the vibrational component the more apparent is the energy that is described.

Reversing Time And Sharing Light

It can easily be seen [OK! It's not so easy when we recognize that we are not talking about our normal "Time-Domain" but instead we are talking about various "frequency domains"] that the adjacent universe would be very different.

An adjacent universe to us would probably seem backwards to us. Even if we somehow changed our "vibration", we could not understand it. We can presume that even time itself would be backwards.

If things are shared in 2 universes, their very makeups would be different. Let's take, for instance, light. As the vibrational patterns come into sync, the photon might become a wave ["have no mass"] in an adjacent universe and be a particle in this one. Think about this. Both universes could share the same light and some nutcase over there would be saying light is sometimes a wave and sometimes a particle. Guess what!

The nutcase might actually be right. Speaking of sharing particles in two universes, we must look at magnetism and electricity or something we now call electro-magnetivity.

Electric-Magnetic Duality

The theory of electro-magnetism is especially interesting. James Clerk Maxwell established the theory, in 1864. The idea that electricity and magnetism are very closely related has initiated many of the discoveries that make up our everyday world, from the propagation of light to the operation of electric motors and generators. The interesting thing is that Electricity and magnetism have a lot of symmetry between their associated fields until you add electric charges. This is because magnetic charges or monopoles haven't been found in this universe. Currently the near symmetry is intriguing, but no reason has been established for the similarities until now. If we determine that electricity and magnetism are dualities, one universe senses magnetism when the other senses electricity and while we find no monopoles or magnetic charges, the associated universe would find no electrical charges. Their circuitry would run on magnetism rather than electricity. Their homes would have magnetism shot through wires to power appliances of all sorts. If we assume that the other universe is not knowledgeable about our universe, they are searching for the elusive electrical charge that cannot be found.

Not So Simple

Of course, the intersection of universes is not so simple as shown in the previous example. There are many universes and each interacts with different "similar" dimensional elements. The graphic following on the left shows how 6 adjacent universes might interact with our current one. I made them look like squares because it is easier to see, but remember that these are all determined by vibrational boundaries and the universes are really governed by vibration rather than distances so you are not looking at sides of cubes but vibrational resonances "sort of".

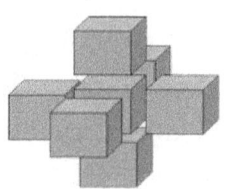 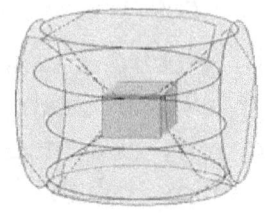

7 interactive Universes   10 interactive Universes

The transfer between any universe and our own would result in a similar confusion as depicted previously. The reference would be different and the confusion would be undeniable. The Bible indicates that there are 7 heavens or universes to consider besides this one [check out Corinthians, Esdras, and Enoch II]. Other religions believe that there are as many as 12 adjacent universes, but we should stick with the 10 identified in the M-Theory right now, because it seems to work mathematically. A potential 10-universe system is depicted above on the right. Each of these universes has several things in common. When put in the time space coordinate system, they all extend out to infinity and yet they all interact.

What I'm trying to say here is that to show interactive universes in space-time is impossible.

To make it even worse; what if you tried to describe 10-dimensions making up these universes? Remember that most of the dimensions are [SUPPOSEDLY] "compactified" according to most scientists. What if these dimensions are compactified with respect to the solid universe, and are simply characterized as something else. What if all 10-dimensions are controlled by some type of vibrational component? Later, I will present a theory that has all 12 dimensions used to define this universe. I think you will agree with me that this makes more sense than having only 4 dimensions and 6 compactified dimensions. Right now, it is just concentrate on the 4 we know modified to be in the frequency domain instead of the time-space domain. That is confusing enough right now. While I'm telling you about vibrating matter and vibrating photons that are sometimes matter, let me go further and tell you just a little about a different concept before we go on.

12 Basic Vibrational Fields

If vibration is the fundamental component of matter, what do you suppose the fundamental component of electricity? [I'm saying vibration in a soft voice as sort of a hint.]---You got it. Electricity is made out of vibration. Now, I'm going to give you a harder one. What is magnetism made out of? Man, you are good. Vibration is the answer again. The list below shows the various building blocks of our existence that have basic building blocks that are simply vibration.

- *Matter –a fermionic vibrational field*
- *Gravity- a gravitational vibration field*
- *Nuclear Force- Fermio-gravitational vibration field*
- *Electricity- an electrical vibration field*
- *Magnetism- a magnetic vibrational field*
- *Photons- Electro-magnetic vibrational field*
- *Lifeforce- Living vibrational field sometimes called "self"*
- *Subconscious- Life Component sometimes called "soul"*
- *Life transfer- Life-consciousness field called "Spirit"*
- *Time-[The effect of vibration on matter, life and energy]*
- *Anti-Time [backwards time]*
- *Space- [A characteristic of time]*

I know some of these are familiar to you: gravity, photons, magnetism etc.; but you are thinking strange thoughts about life and consciousness being simply a vibration. Don't worry about those things right now, because they don't matter when we try to define matter but there is something that should be addressed here. That is something called resonance.

Resonance And The Atom

In electro-magnetics; resonance has been used and studied for years. Simply stated, the resonance thing is the most comfortable frequency for an electric and magnetic field to stay at. It is a point where electric and magnetic fields both have the same strength and when that occurs, the effect of the 2 fields is most noticed. Well, the same thing happens in any vibrational structure. This same feature applies. Let's look at the Aether and gravitational resonance.

Aethereo-Gravitational Resonance

I'm going to get into what makes the various atoms different so listen closely. The difference between a helium atom and a gold atom is vibration, but what keeps the gold atom together? The answer is resonance. Just like the electro-magnetic resonance, particles express this same feature. Particle resonance is the most comfortable frequency for an Aether and gravitational field to stay at. It is a point where Aether and gravitational fields both have the same strength and when that occurs, the effect of the 2 fields is most stable. If these two fields are stable at a high frequency, they appear to be a large atomic mass. At lower frequency resonances, the lighter atoms become apparent.

Life-Consciousness

You might have gathered from the similarities of the various groups of filed that the esoteric components of life and consciousness would also have this resonance feature and so it does. In this case a life force and consciousness level would be matched to provide the most stable life pattern. Those attaining the highest level of consciousness would naturally produce the highest level of living, but this gets into something Indian Gurus call chakras, so I'm going to wait on this level of weirdness for the companion book "the 12-Dimensional World". As we get back to vibrating particles, let's start by looking at the ancient people.

Ancients Used Vibrational Matter

I know that what I have told you so far seems odd, but so does the concepts of string theorists with their black holes and that "Event Horizon" thing where matter is supposed to be made. Just because something sounds strange to you or me does not mean that it is not an accepted idea or theory. It also doesn't mean that it is incorrect. It simply means that the theory is hard to understand. The early humans and a special group called the Nephilim [These were advanced and civilized people that lived before the worldwide flood associated with Noah.] must have known about the make-up and workings of atoms and this whole super-symmetry or parallel universe thing. They also knew and used the concept of vibrational mater. I will call the other universe "heaven" because I'm comfortable with it, but you can call it whatever you like.

Who Were the Nephilim?

Before I go on, let me very quickly tell you a little more about these people called the Nephilim [They are also known as ANAK or Anunnaki.] Many ancient stories tell us that once there was a huge war in heaven. A whole bunch of the angels living there sided with one of the archangels named Gadrael who was later known as Satan. The Satan side lost the war. Punishment for losing was harsh. God took away their "light" [whatever that is]. Losing the "light" made the angels become human. These losers were the ones called Nephilim. The book of Jeremiah [in the Bible] tells us that after the war, all the cities in the world were in shambles and the earth was without form and void. We can be pretty sure when this happened, because the last time the earth was destroyed like that was 65 million years ago. [I know that seems like an inappropriately long time ago. Don't worry about that weirdness in this book. It will be plenty weird by itself.] Certainly, there was another major war about 12 thousand years ago when the planet RAHAB was destroyed, the earth was peppered with hundreds of thousands of meteorites, and giant mammoths were quick frozen with food still in their mouths, but this first war was even worse that the RAHAB War.

Ancients Studied Particle Frequency

Let's get back to our understanding of vibration. These ancient people hadn't been studying these things for the last thousand years, like us. They had been studying these things for tens of thousands or even millions of years. According to ancient texts, they had discovered how to change the frequency of some of these tiny particles. This ALCHEMY was accomplished with tone generators, which could make some of the particles, change their relationship to those particles nearby. In some cases, the new particles would not sense the gravitational pull of the Earth, because the Earth was invisible to it. If that sounds fishy, remember the neutrino and several types of bosons described before. They can go right through the Earth. These ancient engineers could make another tone or electric discharge and a solid object [like a giant stone] could begin to melt into place on a building and spaces between stones would no longer be present just like the incredible spacing of blocks found on PREINCA structures today. These things seem fanciful, but the evidence of this type of manipulation is all around the work in the artifacts from the ancient times.

Vibrational Geneticists

These ancient people were scientifically more advanced that we are today. According to "The Epic of Creation (*Enuma Elish*)" and "Epic of Gilgamesh", the Sumerians tell about the ancient geneticists that created the dinosaurs and eventually a new race of people. The characteristics of primitive animals and people were dramatically changed with substantial knowledge and manipulation of genetic coding. While a major element of this modification certainly could be traced to DNA manipulation, the essence of designing viable entities must require something else. These ancient people must have been able to extract keep elements of life in animals to establish new procreative life forms. This would have meant that the vibrational levels of one species would have been altered to establish a secondary life form. For instance, the DNA structure of a cow and person are almost the same, but a cow cannot procreate with a person. To make the DNA structures identical, the vibrational patterns of the DNA structure would have had to be migrated to various cell groups so that the DNA would generate in a similar way. We don't know how to do this today so no creation of animal types has been accomplished. Certainly, we have yanked out pieces of DNA and placed them back at new locations and have caused all types of mutations, but the ancient texts talk about producing viable animal life.

- *The Igigi, [while these guys were the angels of the Bible, another group of angels became known as the Anunnaki. This group was banished to earth] under the direction of Taimat rebelled against Enlil, and surrounded heaven. [This is the heaven war discussed in the Bible with Taimat taking the Satan position.]*

- *One of the gods, Hubur, created a horned serpent, a mushussu-dragon, a lahmu-hero, an ugallu-demon, a scorpion-man, umu-demons, a fish-man, a <u>bull-man</u>, and others to fight in the war. [Everyone said the same thing. Many monsters besides the Dragon were created especially for the heaven war. The bull-man was like the Minotaur from Greek Mythology and like the half bull half man of the Hittite history.]*

- *Taimat made the dragon to be as a god to fight in the war. [The dragon was not just another pretty face. It was so powerful, this dragon thing was like a god. We will find this identical reference in the Jewish accounts.]*

- *Marduk destroyed Taimat in the heaven war. [Clearly this is the same heaven war that is presented in Jewish history.]*

- *Aruru on the direction of Ea, mixed clay with the blood of one of the other gods to make seven men and seven women to bear the workload of the Igigi. [This indicates that the 6th day man was created by an angel under the direction of God. It also indicates that the human creation was to be a worker for the Igigi/angels. It also confirms the genetic research that these ancient people did. There are literally dozens of ancient texts that confirm this story.]*

84

Mongolian Scientists

The Mongolian Creation of Man story is slightly different. It goes like this.

- *God created a man and woman out of clay. Their entire bodies were covered with a layer of fur.* ***[Most descriptions of primitive people were with hairy bodies.]***

- *This was during a time when the seas were still rising*

- *He went to get some "everlasting life water" and ordered a dog and cat to watch over his new creations.*

- *The devil gave the animals some milk to distract them while he urinated all over the humans.*

- *God was angry with the dog and cat for not caring for the humans.*

- *As punishment, God made the cat lick off all the hair from the human body except for the area around the groin and under the arms.*

- *After each lick, God placed the fur taken away and placed it on the identical area of the disobedient dog.*

OK! This has nothing to do with early man using advanced science, but haven't you wondered why dogs have hair everywhere except the groin and armpit and man has it the other way around? Now you know.

By the way, it also shows that ancient "hairy people" went through some metamorphosis to become much less hairy in almost every culture.

Vibrational Flying

While ancient people could have generated flying machines similar to those produced today, there is ample reason to believe that the hundreds of indications of flying vehicles during ancient times used a different type of propulsion than we use today. The vehicle arose from the ground into the air as if they countered gravity. The only 2 ways to modify gravity is to either approach absolute zero temperature or to change the characteristics of matter. This would have been done by manipulation of the vibrational characteristics of matter. Hundreds of documents talk about the ancient flying ships, while most were from the Sumerians, Babylonians, Jews, and people of India, the fact that the ancient people had made and used flying ships was well known. Once people could manipulate matter by modification of vibrational patterns, the levitation required for these huge flying machines was simple.

- *Peru Tradition-In ancient Peruvian writings, the goddess Orejona landed in a great ship from the sky.*

- *Chippewa Tradition-These American Indians told of the Gin-Gwin [Flying Boats] in their historical tales.*

- *Navaho, Pintes, and Hopi-They all told of the Golden Strangers from the sky that came in flying canoes which were armed with Burning Rays*

- *Brazilian Manacitas Tribe-* One of their cherished legends talks about the Macunbeiros, which were flying wizards that flew inside <u>circular,</u> luminous, machines.

- *Eskimo Tradition-*Their tradition states that they were brought to the north by gods with metal wings.

- *India-* Vimana Flying machines mentioned everywhere. Battles were fought in them and they were used for transport around the globe.

- *Sumeria-* Merkaba flying machines drawn on just about everything; sometimes with several occupants flying overhead.

- *Jewish-* Fiery Chariots described and flying machines with a wheel inside a wheel with portholes around the entire saucer shaped ship.

- *Chinese Dropa-* wrote historical documents about flying in a vehicle and landing in china 12 thousand years ago.

- *Dogon Tribe of Africa-* They describe a flying vessel that came down making jet-like noises and fire coming out of its base before landing.

Weapons by the Mighty

Flying, money building animals and finally nasty weapons, ancient people certainly knew about details of science. Below are a few discussions about their weaponry.

- *"Ramayana" [ancient Indian writings]-"Charged missiles mingled with each other and were surrounded by fiery arrows that covered the Earth and heaven that had increased conflagration. All were scorched by the Brahmastras [Nephilim gods]. They felt the fire that burns the world" [Brahmastra missiles were, evidently, used in the world war.]*

87

- *Generation of Adam 11:3 [Ancient Jewish Gnostic]-* "Leboa, Daughter of Tamar, devised a "Sword of Light" which penetrated the wall of defense around the city of Haner and began to drain the power from the wall." **[Some type of Laser beam apparently penetrated whatever the wall was and drained its "power". Sounds almost like the science fiction of today, but these preflood, world war weapons must have been amazing.]**

- *Etruscan Folklore--Their* traditions indicated that they came from an overseas land that submerged under the sea during a **great war**. **[This indicates that the preflood Atlanteans were probably part of this huge world war struggle and the stories date the great wars to before the flood that sank the great Island Trade Centers.]**

Ancient Space Travel

These guys were so advanced that they had colonized the near planets and naturally, wars were brought to the outposts. Here are a few of the many descriptions. It can be surmised that they clearly understood how to modify mass by manipulation of particle vibration.

- *Indian Details-* "Atlanteans in Vailixi flying ships" and "Indians in Vimana flying ships" **battle on Earth and Moon** as recorded in the "Ramayana".

- *Maharishi Bharadvaya [more Indian details-* In this work there are direct indications of gigantic battles in heaven.

- *Babylon Details-In the* "Epic of Etana" we read, "Etana looked down and saw the Earth had become like a hill and the sea a well and so they flew for an hour and Etana looked down and the Earth was like a grinding stone and the sea like a pot. After the third hour the Earth was only *a speck of dust and the sea no longer seen"* **[The ship, of course, was going into outer space.]**

- ***Chinese Mythology-****Methodology of how to* **send a detachment of men onto any planet** *was described in ancient documents from Lhasa. These documents were found fairly recently and have been only partially deciphered. The remaining information is being deciphered as we speak, so we may find out more about the space war in the near future.*

- ***Greek Legend-****From Greek legends talking about battles between the gods we are told the following: "Hot vapor lapped the titans, flames unspeakable rose bright to the upper air [outer space], lightning blinded their eyes."* **[Apparently lightning weapons were used in outer space]**

On and on I could go, but I think you get the picture. Ancient people understood stuff. One of the things that they knew about was manipulation of materials. They could levitate heavy objects, turn one type of matter into another, and even grow stones into place. All of these things were possible because they understood that particles change characteristics by changing the basic vibrational component of the particles.

Meddling with Heaven

One problem they had and we also have today is that as things were modified here, the other universe had similar modification felt. By many accounts, one of the most disturbing things to the inhabitants of heaven was the manipulation of genetics, which could have required molecular changes. Another thing that probably disturbed the heavenly beings was something called alchemy.

Vibrational Alchemy

I know this sounds like more fantasy, but people who could change matter by affecting the vibrational patterns in the olden days were known as alchemists. There is a substantial amount of written history presented around this whole capability. If we assume a super-symmetry world model, the people in the "other universe" didn't care for those creating and changing particles because they would have "disrupted" the other universe/s and they might have reacted. Some of the reactions are covered in this set of books but right now let's concentrate on alchemists.

The alchemists were real and powerful, at least to some level in our ancient past. They used strange tools for their "Craft". One of the tools of these guys was what is typically called the Alchemic tone. **[Hopefully the tone part is starting to sound familiar.]**

The Alchemic Tone

What I mean by Alchemic tone is that it is becoming more and more obvious by the current studies of the unified particle theorists that tone is the answer to the mysteries of the sub-atomic particle. As I previously presented, the defining part of matter does not come down to the 115 different types of atoms, but comes down to the vibrating frequency of even smaller particles. In fact, the particles don't really exist without the vibration. Instead of using particle structures, scientists have now mapped out various components, elements, and systems by wavelength [association with electro-magnetism], and frequency [association with what becomes gravitation]. We're not just talking about radio waves here. If you can produce the frequency components of a substance, you can modify that substance. This modification could mean levitation of a material or even **making gold out of another substance**. I know this sounds lie hocus pocus, but people are doing this today. Well talk about some of them later.

Enoch

I don't know what this "modification" did in heaven, but let's see what the book of Enoch had to say about knowledge concerning alchemy. [As a note- the authority of the book of Enoch is addressed in our current Bible and was most likely excluded because the texts were not available to the clergy that were trying to establish the "canon" books of the Bible.]

Enoch 64:1-"A commandment has gone out from the Lord that all that dwell on the Earth shall be destroyed; for they know every secret of the angels, every oppressive and secret power of the devils, and every power of sorcery. They know how silver is produced from the dust of the Earth." [Clearly, alchemy is addressed in this verse. It would be a neat trick to make silver from dirt, wouldn't it?]

Event Horizon Confusion

I know this was and still is somewhat confusing, but at least I'm **not** concentrating on the creation of visible particles by close interface with an event horizon along the entrance to a black hole as many of the mathematical scientists are doing today. The mathematicians seem quite comfortable with

91

creation coming from an unseen all powerful, omnipotent black holes, but they seem to be trying to recognize the reaction in space-time. All I'm discussing is how to change one substance into another. It's as simple as making the right vibration. –Nothing to it! Especially if you ignore what you might be doing to our conjoined universe.

Modern Alchemy

Don't think that you will go out and make a tone generator that outputs the 8.5 Exahertz [8,500,000,000,000,000,000 vibrations per second] required to produce Silicon. It can't be produced with existing electronics due to the finite speed of electrons that we can currently pass over any semiconductor material. This type of oscillation could possibly have been manufactured by some means in the very ancient past. This may account for many of the very ancient developments we tend to disbelieve including antigravity, alchemy, and melting stones. There was a reason that the alchemist from the ancient days was feared and worshiped. He could, quite possibly, have been able to make gold out of lead; lift giant stones; and even melt blocks.

Cinnabar On A Stick

Whether he used a large crystal tied to the end of a stick or not, I cannot say, but we do know that crystalline matrices have a characteristic piezo-electricity or vibration after being struck or compressed. Some of the vibrations that can be produced are at very, very, very high frequencies. It is reasonable to believe that this high frequency characteristic would have been most noticeable in crystals of very heavy materials--- let's say cinnabar. I know you are wondering where I got that from. Well, crystalline Mercury [HgS] was well known and mined during ancient times. The name is derived from the Eastern Indian word for "dragon's' blood". Ancient Indian texts are filled with descriptions of some crystals that allowed their ancient vehicles to fly. The crystals were cinnabar along with something called serpent slough. People are going wild trying to find intrinsic properties of Mercury. Let's look at what the very ancient collection of history books called "Mahabharata" had to say.

- *Indra and saw thousands of vimanas [flying vehicles] invented by the Gods lying at rest" The ships were-12 cubits in circumference, had 4 wheels, rose in air, and as they flew, a charge of **mercury** caused roaring flames to shoot out.*

- *Another place it says, "Place a mixture of lode-stone, **mercury**, mica, and serpent-slough on the north and crystals in the center of the engine."*

- *Still another passage says, "Place one type of "mani" in sulfuric acid and another type place with magnetite, mica, **mercury** and, [of course,] serpent-slough. All five **crystals** should be equipped with wires passing through glass tubes. Wires should be placed from the center in all directions, then a triple wheel will set the revolving motion and the two glass balls inside will turn and increase speed, rubbing each other. The resulting friction generates 100-degree power.*

- *Still another passage says, "Place the **mercury** engine with its iron heating apparatus below. By means of the power latent in the **mercury** which sets the driving whirlwind in motion, a man sitting inside may travel great distance in a marvelous manner."*

- *Here's another that talks about an engine, "Four strong **mercury** containers must be built in the interior structure, heated by the controlled fire."*

- *Still more details are provided, "The vimanas develop "thunder power" from the **mercury**."*

93

Like the cinnabar, the serpent-slough stuff is also crystalline so that is going to be even harder to understand.

There is a slight problem in that the output of many of the experiments is a deadly gas. Breathing the stuff kills the experimenters, but at least they found out that cinnabar is piezoelectric. There are about 500 piezoelectric materials known today. The most widely used is Quartz but Cinnabar is quickly making a name for itself. Just like it had done in ancient Greece. In his book "*On Stones* ", Theophrastus of Eresus (371-286 B.C.), a student of Aristotle, described a method to recover mercury from cinnabar by mechanical energy. The metal was obtained from native cinnabar after rubbing it in a brass mortar with a brass pestle in the presence of vinegar. While this doesn't give insight into the vibrational characteristics of Cinnabar, it does show that the crystals were very common. One might believe that a chunk of cinnabar crystal was tied to a stick and the stick was struck to initiate a vibrational sequence that could have done weird stuff. Maybe it could change a person into a newt.

Common Material Frequencies

On the following page is a table that shows the actual or theoretical frequency and wavelength standards of common elements known today. The frequencies have been derived from the various groups investigating the Tamashii model as presented above. How would you like some particles vibrating at 60 exahertz? That vibration causes Gold, as you can see from the list following. Have the right frequency and make the material you want. One thing to note as you look at the table; vibrating frequencies that create gamma rays are also starting to create matter. In this case it is the tiniest block of matter we call hydrogen.

Brief Chart of Vibration Definitions

Name /atomic #	Maximum Wavelength [meters]	Highest Frequency [Hertz]
Brain function	5×10^7	6×10^0 to 10^1
Human hearing	1×10^4	20×10^3
VHF [radio]	1×10^0	30×10^7
UHF [radio]	1×10^{-1}	30×10^8
SHF [radio]	1×10^{-2}	30×10^9
EHF [radio]	1×10^{-3}	30×10^{10}
Microwaves	2.5×10^{-4}	12×10^{12}
Infrared [light]	1×10^{-6}	30×10^{13}
Visible light	4×10^{-7}	75×10^{13}
X-rays	1×10^{-8}	30×10^{15}
Gamma Rays	1×10^{-9}	30×10^{16}
Hydrogen/1	1×10^{-9}	30×10^{16}
Berylium/9	1×10^{-10}	30×10^{17}
Silicon/28	3.5×10^{-11}	8.5×10^{18}
Zirconium/91	1×10^{-11}	30×10^{18}
Gold/197	5×10^{-12}	60×10^{18}
Meitnerium/270	3.7×10^{-12}	27×10^{19}

Magic Crystal

You have probably heard about crystals having some magical power and dismissed it as some type of belief destined to go along with astrology and extracts of poppy seeds. The Tamashii model of atomic structure and this whole concept of vibrating particles may give credence to the notion that crystals hold magic. If you start with a crystal of a homogeneous material that is locked in a covalent lattice structure, it will tend to vibrate at a very specific frequency when excited and the vibrations will continue for some time due to the resonance of the crystalline substrate. In other words, a crystal could cause a continuing vibration. A secondary vibration from a sound cue or other stimulus could very well produce that "beat frequency" I talked about earlier, which possibly could modify the atomic characteristics of a material in close association with the crystal. All this seems like hocus pocus, but that is exactly what causes your crystal watch to work and how a transistor amplifies a signal in your television set. The concept is only modified because the output desired is this extremely high frequency vibration pattern that affects particles.

Caution

Don't discount the magic crystal thing, but don't go out and get a crystal to make you feel better either. It probably will just sit there and do nothing for you. Just open your mind to possibilities that ancient humans could do marvelous things that we are only now beginning to understand. Because they were talented in so exotic ways, the people of that time didn't worship the Nephilim. After all, they learned many of their secrets.

Huge Crystals

While we are talking about crystals, I thought you would like to learn about a neat place that is filled with crystals. I don't know if any special things can happen here, but just look at the size of the Selenite Crystals in the Crystal Caves of Chihuahua, Mexico [see next page]. The person is "normal sized". The crystals were found in April 2000, below a Silver and Zinc mine, 1000 feet down in the limestone host rock. These were just ordinary crystals. They are up to 36 feet long and weigh up to 55 tons, so don't try to put one of these babies around your neck. Oh, by the way; people don't stay too long looking at these marvels because the temperature in the cavern is around 136 degrees. It's still a magical place.

Speaking of crystals, let's investigate a place where crystals could really be compressed to make electricity.

Vibrational Electricity

Here is where we find the "electro" for the electroplating that was done around the world in the ancient times. People needed electricity and there is ample proof that people made electricity out of the piezo-electric effect of crystals. We will also talk about electricity made from differences in electro-negativity of dissimilar metals like our modern batteries, but this first way goes better with the crystal discussion I just had. I must warn you that you are going to think I'm nutty, but the evidence is amazing.

It seems that one of the major developments from before the worldwide flood of Noah was an electric generating plant. I know it will be hard to believe but stay with me for a little before rejecting this notion. Now for the nuts part- we call the electric plant "the Great Pyramid" and we can be sure that it was working before the worldwide flood and was still operational for many thousand years after the flood.

All these supernatural "pyramid power" experiments people do today are missing the boat when it comes to the use of the great pyramid. People place items under cardboard pyramids in hopes of making things last longer and build pyramidic structures to make them feel better, but things still get only and no one really feels better, the reason is simple. The pyramid wasn't some magical structure that was governed by its mystical shape. It, most likely, produced almost free

98

power that could be supplied to remote sites without wires. I know that sounds crazy, so I'm going to have to discuss the theory behind the crazy statement or this amazing element of history will only sound like an absurdity. If it didn't sound absurd at the top of the page, please understand, I'm talking about getting electricity without wires from a bunch of blocks.

I just thought that I'd emphasize the apparent absurdity before we examine the evidence and the modern experiments. Crazy as it sounds; modern experiments <u>strongly</u> suggest this probability.

The reason it's interesting in this book is not so much that electricity was made, but the probability that the electricity was converted to something else for transport. It also shows the basic association of vibrational electricity and vibrational matter. Before we get to the transport let me introduce the Pyramid electric plant.

Real Pyramid Power

Finally, someone came up with a plausible reason for making the Great Pyramid. It explains-

- Why it was made to such exact dimensions. The exactness of the compartments is within millimeters and people somehow believe the thing was dressed with stone tools and sanding stones.

- Why all the different internal chambers were designed in the great pyramid of Giza when all the other were designed with simple corridors to tombs.

- Why 5 granite boulders were necessary when no other pyramid uses anything like this.

- Why the granite slabs were polished where they could not be seen,

- Why the creators made the tiny, sub meter sized shafts that went through solid stones at angles that were extremely difficult to manufacture.

- Why the thing was made in the first place when it NEVER had a body placed in its belly.

Christopher Dunn and Electricity

Christopher Dunn initially theorized the concept of a pyramid electric generator and has tested his generator theory with very positive results. His theory states that the great pyramid could, very easily, have been used to produce large amounts of electric energy. My version is somewhat simplified, but please believe me when I tell you that the occurrences of all of the anomalous elements associated with the Great Pyramid were not happenstance. The ancient people went to a lot of trouble to make the great pyramid the way it was. Oh yes, don't believe anyone telling you that the Great pyramid was made a mere 6 thousand years ago. While I'm not going to get into it in this book, we can be pretty sure the thing has been around for the last 40 thousand years. Much of that time it made electricity.

The Great Pyramid is loaded with unbelievably accurate shafts cut into and through limestone at oblique angles and granite slabs are critically placed and polished where the polishing cannot be seen. Added to this strangeness was the discovery that the chamber dimensions show remarkable resonance levels, which we will address in a minute. Recently we found that at least one of the miniature shafts had been sealed up under the orders of some ancient pharaoh. The great pyramid also has five, high crystal content, polished, granite slabs positioned on top of one another but spaced apart so that they could vibrate and below them is located a very interesting chamber that shows the signs of explosions. All of these unusual features were put into the pyramid when it was built, because it was built for a purpose different than holding a dead body.

Production of Gas and Salt

Today we know that mixing an "acid" and just about any "base" together forms hydrogen gas. We also know that pure hydrogen is extremely volatile and will cause an explosion if it comes in contact with air and a small amount of heat or spark. After the reaction is completed, the hydrogen would have mixed with oxygen in the air and would have produced water. The remains of the acid/base mixture that was left after the gas was produced would have turned into a salt.

Did I mention that salt deposits have been found in three places in the great pyramid? Large quantities have been found in the "queen's [gas production] chamber", and smaller quantities have been found in both the "grand galley [gas baffle]" and the "king's [resonating] chamber".

Electricity From a Crystal

Let me get back to some basics again. Today we constantly use the fact that when quartz crystals, such as those found in granite, are compressed, they produce electric sparks. This is called the Piezo-Electric Effect. When the pressure is relieved, the crystals produce more electricity. We also use the fact that a crystal will deform itself at a specific cyclic rate that is dependent on the cut of the crystal and its dimensions whenever excited by electricity. [sort of- piezo-electricity in reverse] I'm not talking about the whole crystal power thing and the act of placing a crystal on your body to change your mood, relationships, or wealth. That stuff seems somewhat crazy to me but the power in all crystalline structures noted above has been used in electronic equipment for many years.

Electrical System

It looks like the very ancient Egyptians knew these things and made a gas manufacturing, granite crystal compressing, electricity producing, machine which we call the Great Pyramid. A schematic of the pyramid as an electricity generator, which shows the component parts, is pictured below. The names given are not the same names we have been told, but these are most likely the functional names.

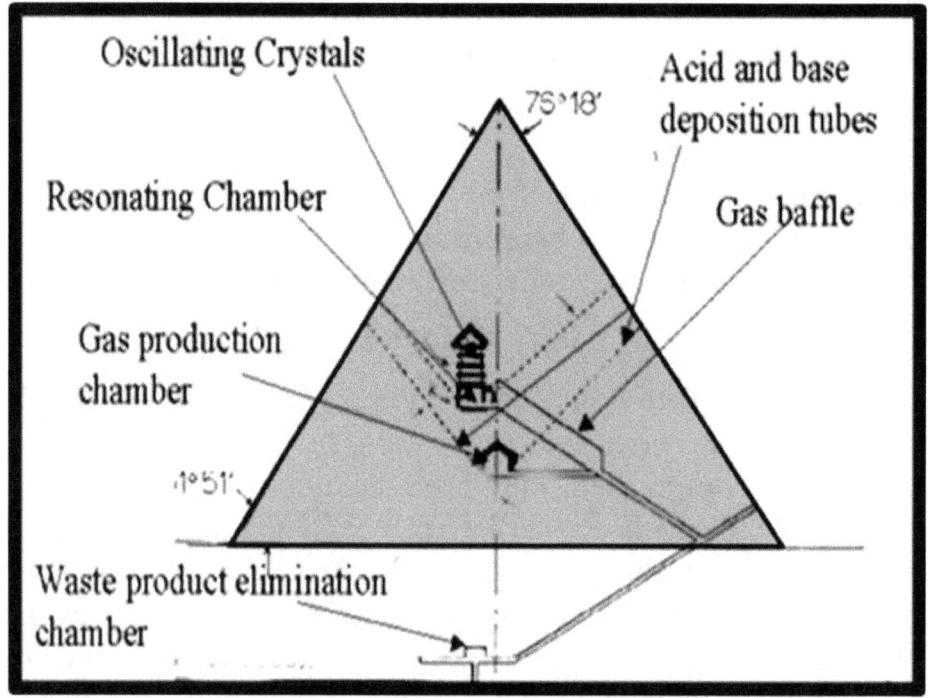

Resonating Chamber and Oscillating Crystals

As discovered recently by researchers Rock McCullum and Bill Cox, almost everything in the "King's Chamber" resonates at 640 Hertz. In the "king's Chamber, five polished red granite crystals are placed above the "king's chamber and are separated from one another. My bet is that the crystalline mass making up each granite block above the "King's

Chamber" has a resonance of 640 hertz. The granite blocks were not just found to be 640 hertz, they were "tuned" like we tune crystals today, by polishing them down to the precise dimension required. A close-up of red granite would show high quartz crystal content that could be compressed to form electricity.

 Some would look at the massive granite oscillators shown in the drawing below and note that if only one side was polished, they would not be efficient at producing electricity and they would be right. They would also say that granite is not a very good crystal, but it does contain the piezoelectric quartz crystals and if you get enough granite and press it hard enough, you get a lot of electricity. The polishing was only used to change the resonant frequency of the giant crystals.

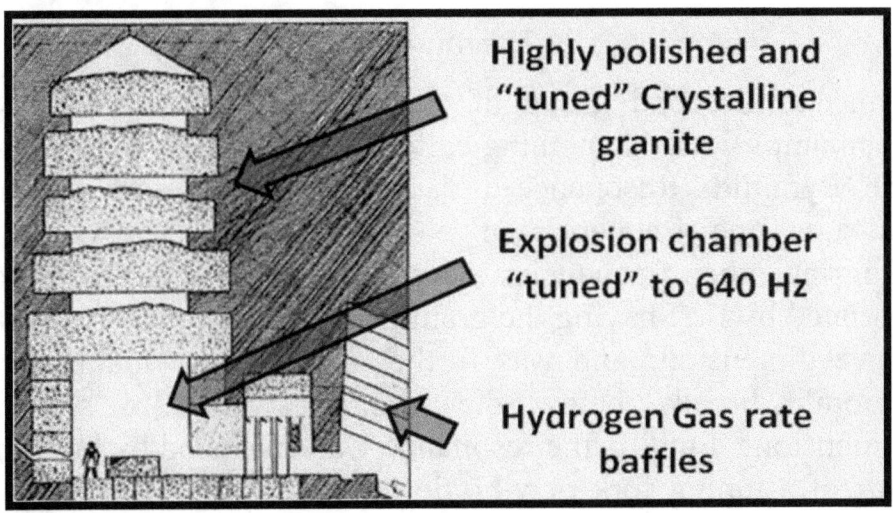

Highly polished and "tuned" Crystalline granite

Explosion chamber "tuned" to 640 Hz

Hydrogen Gas rate baffles

Once the chambers, shafts, resonating crystals and properly sized resonance cavities were in place, all the Egyptians had to do, was find a way to continuously compress the crystals enough to produce electricity. Someone though about it for a little and might have said, "The pyramid will compress the granite by itself." Sure enough, the weight of the pyramid pushing on the granite caused electric energy to be output from the crystal. It wasn't some magic and they could not have kept the crystals from producing the initial surge of electricity, even if they wanted to.

The Instantaneous Electricity Was Now Gone

From pressure on their structure, the crystal slabs, like all crystals would resonate and begin to compress and expand itself at its base frequency. Typically, these "oscillations" die down very quickly, but simply having this great pressure on the blocks would produce several "cycles" of electricity. Of course, the electricity had nowhere to go so a very high "voltage level" would be reached and like miniature lightning bolts, electrical sparks would be generated. Then the energy would dissipate as heat and go away. I know this sounds like it was useless even with resonance.

Resonators

This "millisecond group of electrical sparks" would have been amplified if something called "resonance" was added in the Pyramid. Resonance is a property that amplifies or sustains one frequency of electromagnetic energy while ignoring other frequencies. This property of resonance was initiated by first having the granite slabs polished to the same basic dimensions and was further extended by making the "room" directly below the granite crystals to specific dimensions. Finally, the resonance was enhanced by building sort of a tuning fork or vibration cavity at the center of the room. Some may tell you that this vibration cavity is a sarcophagus, but it doesn't make sense. There was never a lid and no evidence that a body had ever been placed in the tuning fork. By the way, the frequency that all of these elements seem to amplify is 640 Hertz.

As shown above, even the lip of the "vibration cavity" was possibly rounded smooth to insure optimum sustainment of oscillation. This is no sarcophagus. Why in the world would the sides have rounded tops so that a non-existent lid could never stay secure??

Gas to Restart the Pyramid

All that was fine if you only wanted one burst of electricity, but sustainment was certainly needed and it was also planned for. Hydrogen gas was, most likely, produced in what many call the queen's chamber by combinations of materials transferred down the small deposition shafts visible today. Some "acid" would be poured down one shaft and some "hydrated base" went down the other. The volatile gases [hydrogen] produced by the mixture would slowly rise through the baffles [sometimes called the Grand Gallery] and finally reach the "King's chamber" before the "Granite Crystal Oscillations" stopped. Sparks associated with the electricity produced by the granite crystals would ignite the gases and cause the chamber to momentarily get larger from the explosion. This action relieved the pressure on the Granite slabs, which, in turn, produced more cycles of electricity. Very quickly the pressure of the pyramidic weight would take over and begin to crush the granite slabs once again to produce more cycles of electricity over again. Like an almost imperceptibly moving engine, the electricity would continue to be produced as long as hydrogen gas was allowed to enter the "Resonating Chamber". I don't mean small

105

amounts of electricity either. It produced large amounts of electricity. I could talk about the getting bigger, getting smaller, getting bigger, getting smaller reaction and the explosions and compressions kept occurring over and over again at 640Hertz, but I won't because it would be very boring.

Explosion Evidence

Researchers have discovered that many of the boulders that make up the king's chamber have been moved out slightly and one of the granite slabs has even cracked due to the effects of these cyclic explosions so there is good evidence to support the oscillating cavity theory.

Black Ceiling

Did I mention that the bottom of the lowest granite ceiling slabs is covered with a fine black dust as if some type of high rate burning process had blackened it? This blackening is found nowhere else in the pyramid. Someone might think explosions occurred in the "King's Chamber".

Starting the Pyramid

After the "queen's chamber" had been filled with the acid and base mixture, and the hydrogen gas had tunneled through the grand gallery to the king's chamber; a flame was introduced through one of the openings to the chamber and the ensuing explosion began the production of electricity.

Sustainment

All the Egyptians had to do was to periodically refill the materials for the gas production and the machine continued to output electricity at about 640 Hertz. You might wonder, "What was so special about 640 hertz?" Nikola Tesla may have discovered one part of the answer.

Tesla Vibrates

This explode, compress, explode thing done by the great Electric-Generating-Pyramid made something we call alternating electric current. Just to make things interesting, let me tell you something odd about alternating current. There is no direct exchange of electrons through a material so there is no electronegative potential that can cause work to be done. It makes absolutely no sense that having electricity go back and forth in a wire could make motors run or electronics electronate.

After saying, that let me tell you the alternating current does work. It works by momentarily energizing a structure such that it can make atoms get larger and then reducing the size of the atoms. This vibration causes the magnetic fields of the structure to materials to build and the collapse of the magnetic fields allow work to be accomplished. It almost like magic, but we take it for granted every day. The master of this unbelievable secret was Nikola Tesla who indicated that its discovery magically came to him.

Not only did Tesla discover alternating current, but he also discovered that **the ground could carry electricity**.

Besides Telsa, no one in recent history has been able to use the ground to carry electricity.

The ground has moisture in it and metals of all types. When electric current is pushed through the ground, the ground simply heats up and no electricity is transferred. In Telsa's configuration, no electric wires were needed to carry electricity as most people believe must be required today. He demonstrated to many this new capability and, reportedly, lit large quantities of electric lamps, which were placed at long distances away from his generator. He, somehow, used the earth as the conductor and lit them without wires. The problem with his newfound distribution method was that no one could control who was using the energy, therefore, no Earth transfer systems were ever commercially produced. No one would gain a profit and Tesla's method died with him.

Tesla's typical experimentation involved substantial high voltage discharges as shown on the next page, but he really wanted to do something special for everyone in the world. He wanted to make free electricity for everyone. Unfortunately, he could not get the backing needed to further his experiments in this area. He had received financial backing from J. Pierpont Morgan of $150,000 to build a radio transmitter for signaling Europe, but later he wrote that he had much larger plans. With the first portion of the money he obtained 200 acres of land at Shoreham, Long Island and built an enormous tower 187-foot-tall topped with a 55 ton,

68-foot metal dome. This was to be the beginnings of an electrical distribution unit that used the earth as a conductor. He was unable to complete his experiment, probably because there would be no money in it for J.P. Morgan who withdrew funding. In fact, everyone abandoned the conquest; so today, we must pay for our electricity. I know that you are probably saying to yourself that conducting electricity through the earth is impossible, but now we know it is not impossible. It's not simple, but it is possible.

A New Automobile

Later, Telsa was said to have used his free air transmission of electricity to power an automobile. After showing how the car could operate without the need for gasoline, these experiments were also defunded and abandoned before Tesla died. The 1931 Pierce-Arrow shown below was supposedly the type converted to run on the unknown energy source. Possibly he had found a way to create electricity by manipulation of the vibrational components in free air.

Possibly, The Engine Used Earth Conduction.

Confirmed Earth Conduction

It really wasn't until 1976 that Dr. Andrija Puharich was able to point out that Tesla's power transmission system could not be explained by the laws of classical electrodynamics, but, rather, in terms of relativistic transformations in high energy fields. It seems that when an electron encounters a positron, the two particles would annihilate each other. Because energy can't be destroyed, the particles are transformed into an electromagnetic wave. Transversely, if a huge electromagnetic wave is manufactured, electrons and positrons of equal quantity must be manufactured. This recombination can be done at a remote site. Let me put this whole thing in the frequency domain and associate it with Vibrational matter.

All one would have to do is to somehow make the electricity go out of phase with this universe and it would no longer exist in this universe. At a destination, simply relink the electricity to this universe by converting the phase back to what it should be.

Ordinary electrical currents do not travel far through the earth. Dirt has a high resistance to electricity and quickly turns currents into heat energy that would be wasted. With this "pair- production" method, electricity can be moved from one point to another without really having to push the physical particle through the earth. The transmitting source would create a strong field, and a particle would only be created at the receiver by means of some catalyst. Without the catalyst, there would be no loss of energy and no heat would be generated. Put another way, the earth could sort of store the electricity until needed.

Evidently, the ground current distribution of electricity during the ancient times was used for many years and anyone that held the secret to the recombination process could tie into the system and pull out electricity without wires. Limiting the number of people who held the "secret" of reinstituting it must have controlled payment for the electricity. The famous Moses probably learned the secret while he was in Egypt. After all, he would have been integrated with the council of magicians in Egypt because of his high placement in the royal family. By all accounts, the magicians were the caretakers of the electricity 5000 years ago. Later, Moses, apparently, used it to the advantage of the Jewish people. It possibly was part of the power behind the Arc-of-the-Covenant that will be reviewed later. Don't get me wrong, the Arc was an unbelievably important and mysterious device in its own right, but it might have been electric.

Pyramid Resonating Dimension

Like the resonance of the "king's chamber", researchers have determined that the base length of the Great Pyramid equals the distance that a sound wave at 640 hertz would travel in one cycle. This critical dimension, probably allowed the transfer of the 640-hertz AC electricity from the pyramid. Its resonance may have been very important in the transfer of energy through the earth just like Tesla had demonstrated in the form of a frequency that was a critical harmonic of the earth's normal vibration frequency. This possibly allowed transfer to the areas around the world that needed the commodity and that is where rows and rows of stones may have come in.

Uses for the Electricity

The ancient dwellers that lived in Northern Europe and the others may have been producing electricity from the generated field of the Egyptians thousands of miles away, so you would think that the Egyptians also would have converted and reconverted electricity for various uses. There is evidence to suggest that the Egyptians used the electricity for lighting, electroplating, and operating some of the tools needed to machine stones. It also was probably used to power something we call Dendera tubes. The Dendera tubes, electroplating, and lighting will all be discussed later in the book, but one electrical component, commonly known is a field of monoliths. Around the world we find more and more of these monoliths. Each of the thin stone blocks is lined up in a peculiar array and we find the same key elements repeatedly. This has to do with Vibration Engineering.

Acoustic Engineering

Huge cashes of upright stones have been found around the world patterned in unusual ways like the ones shown at Carnac [to the left]. Initially they were thought to have something to do with burials, but the bodies were missing. The newest thought is that the upright stones are critically placed to amplify sounds, but what if an area could be "resonated" to accept the 640-hertz signals better. That's where investigator R.G. Jahn Waller comes in.

He has taken sound generators and meters into the central areas of six ancient "multiple monolith" structures and measured their acoustical properties. The sites selected were: Wayland's Smithy, Chun Quoit, and Cairn Euny, all in the U.K.; Newgrange, Cairns L, Cairns I, and Carbane West, all in Ireland. His findings may serve to identify a similarity of many of the sites including the one found in Corsica shown to the right. Guess what? All of these sites date back to about 6 thousand years ago.

On the left is a field in Callanish and still another on the Isle of Lewis is shown on the right.

Mr. Waller found that the central area would amplify frequencies of about <u>100 Hertz</u>. Mr. Waller suggested that this amplification was used by human voices, but the range is too low. What may have slipped by is that vibrations move more swiftly in the ground. While I don't have the test data, we know that it would be significantly higher than the 100 Hz tested. What if the vibrational resonance from the ground is 640Hz? That would clear up many of these mysteries. Possibly the stones were put into places to allow the transfer of electricity more readily. One thing is for sure, the monoliths are for something and they are all spaced a particular distance from each other. In this vein of vibrational elements changing particles to do work for you, there are potentially thousands of different things the vibrational field could have been used for. While most of these things are found in and around the UK, around the world, the whole put an arrangement of vibrating stones seemed to have gotten everyone going with something.

Next are some of the huge numbers of these potential electricity receivers.

Brazilian Field

Tibetan Field

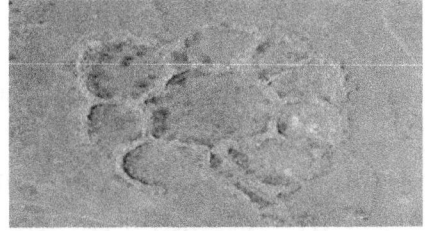

South African Field

Israel Field

Evidence Of Electricity

Before we can go on, we need to investigate more about electricity in general because you are possibly thinking that electricity was the last thing that these primitive people would have had. Not only did early mankind in South America, Babylon, and Egypt all know how to do electroplating of antimony, gold, and silver, but also, they had a great understanding of electricity in general. Two of the now famous electric artifacts are shown below. One is now called the Baghdad Battery. It produces a continuous discharge of electrons as chemicals change their characteristics rather than the alternating current produced by pyramids and similar things but the idea of changing the characteristics of the material and forcing the electrons to leave is accomplished by the same general feature that could cause lead to turn into gold by making electrons and protons leave a structure so don't just think of a battery as something that makes your radio turn on. It is transforming its "matter" right before your eyes.

Geode Battery

The picture to the left [on the next page is some type of power conversion device found **inside** a geode, in California. Below the geode is a drawing of x-rays of the geode showing the elemental parts. These include a spring, core, plate, and electrical insulator. The same parts as you would expect in a battery. Maybe this is a new way to package batteries, but it takes a long time to complete the package as geodes can take millions of years to form. The object is extremely ancient and certainly was manufactured well before we originally thought that everyone used electricity. The central metal core surrounded by the white material looks like a battery, although some even attribute the structure to that of a spark plug, however, I have never seen a spark plug that had a body that was a wide as that shown in the x-rayed geode. Whatever it was, it was electrical. On the right is a drawing of the parts and a size comparison to a standard D-cell battery.

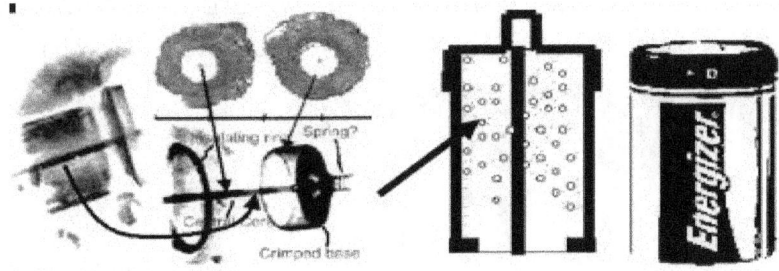

Electrical Connector In Stone

Here is a good one from New Mexico that was found in 2005. A mystery rock with an embedded or implanted, man-made-like, electrical component has come into the possession of a Mr. John Williams who currently lives in New Mexico.

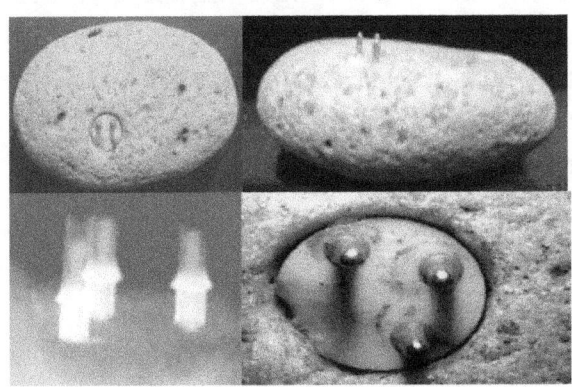

117

The Embedded component is round and about 1/4" diameter. Rods or pins are of steel-like appearance. Component base or matrix appears to be a ceramic or rock-like material. This thing looks exactly like a modern high voltage connector. The preceding picture shows this amazing thing and X-rays show that the "steel" pins are swaged into some type of ceramic base. This assures us the object in the rock is man-made. Unfortunately, the rock has to be millions of years old so some scientists don't like seeing it.

Baghdad Batteries

The drawing below is one a number of ancient batteries found in Baghdad. [A cut-away of the 2.5 to 3-inch jar shows the inside rod, the inner cylinder and where the wires were attached]. Guess what! More of these batteries were found.

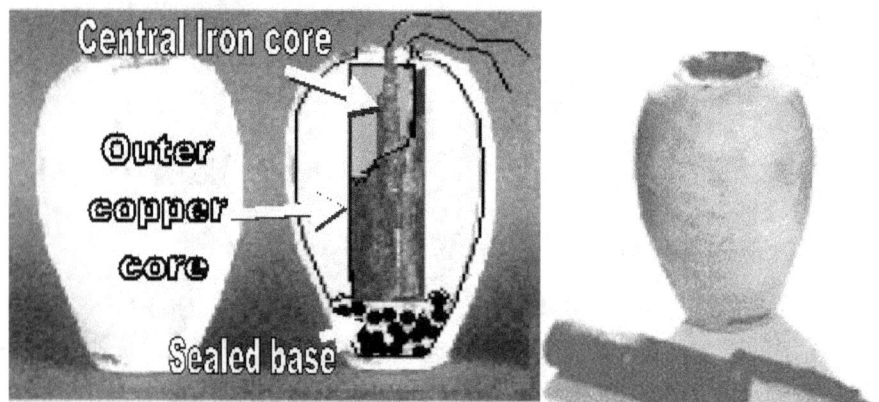

After the initial find of a single battery, 4 more "batteries were found in a magician's hut in Seleucia and 10 more were evidently found at some nearby site and placed on display in a German museum. It seems like these "double-D-cell" sized batteries were everywhere.

The picture below left shows the internal metal cores of an actual Baghdad battery along with a group that are sealed.

To the right shows the battery compared with a common D-cell we used today in place of these older models.

This "Baghdad Battery" contained something else equally as interesting. The metal parts were put together with a substance that has been analyzed and determined to consist of 60 percent tin and 40 percent lead. Anyone in the manufacturing business knows that what they found on these parts was our present-day mixture used as solder which attaches electronic components to circuit cards.

The question that should be asked is what did they use batteries for?

Indian Batteries

If that's not enough, an ancient Indian work named "Agastya Samhita" tells about construction of still a different type of battery. [See below]

Where the Baghdad battery was a Copper-Iron battery, this one is a Copper-Zinc battery. Not only does this show that batteries were available, but also that they were common enough for various types to be available on the market. Let's read what this ancient document had to say. Some of the words have been translated into more modern English while others have been left as the original wording for fun.

Agastya Samhita

- *Place a well-cleaned copper plate in an Earthenware vessel.*

- *Cover it first by copper sulfate and then by moist sawdust.*

- *After that put a mercury-amalgamated-zinc sheet on top of the sawdust to avoid polarization.*

- *The contact will produce an energy known by the twin name of Mitra-Varuna [ancient name for electrical energy].*

- *Water will split by this current into Pranavaya and Udanavayu [Oxygen and Hydrogen].*

- *A chain of **one hundred jars** is said to give a very active and effective force.*

If we assume this to be a normal carbon-zinc reaction, each cell would produce about 1.2 volts so an attachment of 100 jars in series could provide a voltage differential of 120 Volts. Isn't that interesting that the voltage we use today for everything is the voltage recommended in the olden days?

Other Electric Evidence

The batteries and description for making batteries are not the only evidence. Early texts talk about lamps that stay one for years, and other evidence shows electroplating, and motors. Here are some examples. Please note that electricity was used around the world. While some of the descriptions are probable exaggerations, the sheer number of claims reduces the possibility of the evidence not being true and the physical

120

evidence seems to strongly point to artificial lighting and use of electricity of plating.

Roman Electricity

Noma Pompilius- *[second king of Rome] had a perpetual light shining from a dome. Plurtach wrote about it and the priests told him that the light had remained for centuries.*

Pausanias- *This Roman writer talked about a beautiful golden lamp in the temple of Minerva that could burn for years. [2nd century AD]*

Roman excavation- *According to reports in 1401, the burial tomb of Pallus was opened and a lamp was still burning and had been supposedly burning for 2000 years.*

Greek Electricity

Lucian- *This Greek Satirist from around 150AD wrote that in Hierapolis, Syria he saw a shining jewel that illuminated the temple at night and that glowing "Stones" kept light shining all night at Baalbek.*

Later European Electricity

Saint Augustine- *He wrote about a wonder lamp in Egypt- "Neither wind nor could rain extinguish it." [~400AD]*

Oedipus Aegyptiacus- *This document written in 1652 refers to lighted lamps found in subterranean vaults in Memphis.*

France- *Beautiful prehistoric cave paintings were done without any sign of torch light- no blacking of the ceiling or anywhere.*

Hopi Indian Electricity

Hopi Indians- *They used a Light generator made from a rectangular piece of luminescent quartz and a bolster shaped upper piece worked by rapidly rubbing the crystal, according to researcher David Chandler. These lighting machines have been found in several locations in southwest America.*

Incan Electricity

Peru- Gold and silver electroplating over copper required electricity and the low ceilings associated with Chimu buildings had no signs of soot on the ceilings or anywhere which means another form of light must have been used rather than torches.

Egyptian Motor

Some people believe that the following images might represent the schematic diagram of rotor and stator components of a motor. Whatever they are, they look like possibly some kind of schematic or electronic drawing. Wiring or electrical flux lines are shown emanating away from the Device in the center. This type of machine might have been only a small sample of what the electricity allowed the Egyptians to do, but batteries were not big enough the produce large amounts of electricity, and transfer of "Direct Current" Electricity over any distance is impractical, so the Great Pyramid was the major electricity producing plant and these other devices were for small jobs.

Vibrating in History

As it turns out making electricity is one of the easier things to accept from our distant past. I'm going to go over some pretty odd things right now. For years you have simply thought of these things as magic, or lies, or ANOMALIES. Well it could very well be that the Arc of the Covenant, Moses magical staff, the Egyptian Dendera tubes, turning sticks to snakes, levitating blocks, and other things that keeps showing up in our history are true. What was happening with all of these varying types of mystery things is that vibrations produced by some controlling medium converted matter just like the photon changing from visible light to the deadly gamma rays. If one can control matter in some minor way, another thing that could be increased is life itself by causing the DNA or structure of life to be modified. As we confront the possibilities of Matter being controlled by vibration, many of the things that seemed to be fantasy or some religious exaggeration were actually the truth. You wondered how Peter walked on the water and how Elisha's bones brought a boy back to life or how the Red Sea was raised from its river bed. Some simply say God did it. It's magic. Well God doesn't work like that. He created this world with a set of physical laws and the things he or any of his creations do in this world work within those laws. We are only now beginning to understand what some of those laws really are. The most important is the law of vibrational matter. The bringing people back to life one will have to wait until we get into the last of the 12 dimensions so don't worry about that one just yet.

Moses' Vibration Staff

Speaking of Moses' Staff, let's talk about several staffs that were used in what we consider to be magical ways. Here again, I need to use the electricity model. The staffs somehow collected the electricity and used it for required tasks. One of the tasks seemed to be making tones to change atomic structure. One of the obvious things that occur from this is levitation. In ancient text, we are told of sounds or vibrations coming from one of these staffs, which eventually made heavy stones weightless. All of it sounds like some crazy talk, I know, but the alternative is to discount hundreds of ancient documents, drawings, verbal histories, and even claims from the Bible which rely on some magical quality of several magical staffs during ancient times.

Priests of Hike and Thoth

Thoth was the first of a long line of Egyptian magicians called the Priests of Hike and he had one of these staffs. In the Emerald Texts, it was written that Thoth left Atlantis and survived the flood to teach the Egyptians. He was probably one of the Nephalim and he began teaching the Egyptians about writing and all types of what we think of as magic and science today. He was revered as a god, was the thirteenth ruler of Egypt, and was probably responsible for teaching man "magic tricks" and making some of the machines that could perform the magic that are discussed below.

The Mattah Staff

Thoth may have introduced the magic staff [also called the Mattah or UAS by the Egyptians]. It is, by far, the most used and best known "magic device". The power in many of these things was, most likely, awesome. One of the "tricks" it was known for was the ability to "open a lake or river". Not only did Moses have great power when using this device, but other historical records show the strength of the "Staff" when used by other magicians before Moses had his opportunity. Some of the better-known magicians that were able to perform amazing feats, aided by a staff, besides Thoth, are the Hike priests, Bimater, Mises, Jajamankh, Elijah, Elisha, Moses, Aaron, Joshua, and possibly a even a non-fictitious Merlin. All of these people used "staffs" effectively to do things that are thought to be physically impossible to do today. As a note of interest, Moses staff was made from wood, but the common method for building a UAS in ancient Egypt was to allow a bull's penis to dry in a "Stick-like" shape. It was considered to bring power to its owner.

Staff Levitated Stones

One of the more important priest tricks was to move the huge stones by levitation. This was discussed before in the section dealing with construction techniques. According to ancient texts, this was done by using one of the magic staffs to produce some kind of tones that would allow the stone to be lifted and carried through the air for a short distance. I have no idea if the stick had a cinnabar crystal on it or some other generator, but we are certainly told the sticks could transform matter just like electricity does only in a more exotic way. From the looks of the many huge stones used as building material, this levitating thing was almost common knowledge by the "Magicians" after the flood. The ancient descriptions indicating that the staff would cause some type of tones that somehow raise objects off the ground goes along with the ancient Indian texts which discuss levitation and flying machines had something to do with sound.

Staff Levitated Water

The staffs didn't just levitate rocks. They also levitated water. The first story written about levitating water was during the time of king Seneferu of Egypt. He had lost a trinket in a lake accidentally and asked one of the Hike Priests named Jajamankh to get it back. The magician simply took his staff, held it up and the water immediately rose up so that he could walk to the center and retrieve the trinket. Upon lowering his staff, the water filled its normal position. This happened about 2000BC. Here are some of the "magic staff" stories that have survived. Note the Mandaean version of Moses' escape from the Egyptians. That story provides us with a possibility that another river was raised, but during that act a mistake occurred that costs hundreds of lives. Although different from the Biblical text, the Mandaean similarities establish consistency in historical record.

- *Orphic Hymn to Bacchus*- *According to this Babylonian work, the* **sorcerer** *named Bimater had* **a rod** *with which he could work miracles very similar to the one that Moses had. The "miracles" including parting waters of the river Orantes, the river Hydastus, and the Red Sea.* **[This same man also struck a rock and produced water and had his rod turn into a serpent.]**

- *2 Kings 2:8- And Elijah took his mantle, and wrapped it together, and smote the waters, and they were divided hither and thither, so that they two went over on dry ground.* **[Elijah parted the waters.]**

- *2 Kings 2:13-14- He [Elisha] took up also the mantle of Elijah that fell from him, and went back, and stood by the*

126

bank of Jordan; And he took the mantle of Elijah that fell from him, and smote the waters, and said, Where is the LORD God of Elijah? And when he also had smitten the waters, they parted hither and thither: and Elisha went over. *[Elisha parted the waters.]*

- *Josephus-* Moses had thus addressed himself to God, he smote the sea with his rod, which parted asunder at the stroke, and receiving those waters into itself, left the ground dry, as a road and a place of flight for the Hebrews. *[Moses parted the waters according to secular history.]*

- *Exodus 14:21-* And Moses stretched out his hand over the sea; and the LORD caused the sea to go back by a strong east wind all that night, and made the sea dry land, and the waters were divided. And the children of Israel went into the midst of the sea upon the dry ground: and the waters were a wall unto them on their right hand, and on their left. *[Moses parted the waters according to Biblical history.]*

- *Mandaean tradition [Iran]-* Most Jews worshiped Ruha [Lilith?] and knew nothing of the "Light" or the "children of the Light". The Jews tried to escape the Egyptians. Upon their escape, one of them named Musa [Moses] had a staff, and knowledge of secret names. The staff had been given to him by Ruha [Lilith?] and opened into two parts. When they reached the Sea of Suf [Red Sea], Musa took the rod and struck the water and uttered names and the water became solid like ground. The Egyptian followed and the magician **Para Malka** used his Marghna [staff] and also made dry land. The magician continued across the sea while the Egyptians came across. When the magician reached the far shore, the sea

127

turned back into water and all behind him was drowned.
[Malka and Musa both parted the waters.]

Moses was the most well-known Jewish Magician, but he was far from the only one. Jannes and Jambres were also Jewish and well known for their practices, but they were not doing their trade for God. These types of people were known as sorcerers and the Adamics were told to STAY AWAY from their magic. Other great magicians are also noted in the texts below. I use the word great only as a reference to magic capability, not as some people to necessarily be respected.

Moses Staff

Moses was most likely a Hike Priest and was aware of many of the magical incantations common in that day. This did not only include levitating water out of its bank or movement of huge blocks, but also included turning the Mattah [staff] into various animals. Some drawings are shown below of the staff.

The ancient rulers and gods to show royalty did not hold up the Mattah; these people held it because it helped protect them. Certainly, the later rulers held the staff because the earlier rulers had done it, without understanding why, but initially it was a powerful tool. Notice, that on top of the stick, there is some type of long thing [possibly a crystal].

Lotapes, Jannes, & Jambres

Just because a magician was Jewish did not mean that the magicians was necessarily good. Two Jewish magicians [sometimes called sorcerers] did everything that Moses had done to mock his magic. Their names were Jannes and Jambres and they were famous for their magic. They made snakes, frogs and everything else. They probably used a Mattah just like Moses did when he initiated the plagues on Egypt.

- *Jannes and Jambres- And in the presence of the King [of Egypt], he [Jannes] opposed Moses and his brother Aaron by doing everything they had done.*

- *Targun [Exodus 1:15]- Then Jannes and Jambres, the chief wizards, spoke up and said to Pharaoh---*

- *Gospel of Nicodemus- They were servants of Pharaoh, Jannes and Jambres, and they also did signs not a few of which Moses did, and the Egyptians held them as gods. [This shows that the Jewish magicians Jannes and Jambres were not only magicians, but they were also very highly placed in the Royal family.]*

- *Historia Naturalis- There is another magical group deriving from Moses, Jannes, Lotapes, and the Jews, but many thousands of years after Zoroaster----.*

- *Numerius- Next are Jannes and Jambres, Egyptian sacred scribes, men judged to be inferior to none in magic, when the Jews were expelled from Egypt, they were chosen by the people of Egypt to stand up to Moses, the leader of the Jews, and a man of most powerful prayer.*

- *Chronicles of Moses- And after Moses and Aaron left, Pharaoh sent and called to Balaam the magician and Jannes and Jambres his son the sorcerers.*

Other "Staff" Magic

In the hands of Moses, the staff became an awesome weapon. Not only could Moses' Rod aid in levitating water, but it could also perform other miracles including the following: [Although God probably helped Moses a little, it is very apparent from other texts that the "magic staffs" of the day could do many things.

1. *Turning into a snake-* The **Sumerians, Egyptian** magicians [Jannes and Jambres], and Moses were all able to accomplish this feat while holding a staff.

2. *Manufacturing water when the ground was struck-* The **Sumerians** and Moses accomplished this feat while holding a staff.

3. *Turning water into blood-* The Egyptians [Jannes and Jambres] and Moses both were able to do this while holding a staff.

4. *Turning bad water into clean drinking water*. Only Moses did this feat while holding a staff.

5. *Bringing huge quantities of Frogs-*The **Egyptians** [Jannes and Jambres] and Moses were able to do this while holding a staff.

6. *While holding his staff, he was also able to bring locust, darkness, and many other miracles including death.* Normal Egyptian magicians could not do these other miracles no matter how they twisted their sticks. The reason Moses could do more than the other magicians was that Moses' rod was aided by the power of God, according to historical record.

Here are some more magical things that were attributed to the "staff" and written down in historical records. Some of the magic was done by Moses and some was done by other magicians of the day.

- **Egyptian Emerald Tablet-** *Then I [Thoth] raised my staff and directed a <u>ray of vibration striking them in their tracks</u>.*

- **Exodus 7:20-** *And Moses lifted up the rod, and smote the waters that were in the river and all the waters that were in the river were <u>turned to blood</u>.*

- **On Baptism 9-** *Similarly, <u>water was healed of its bitterness</u> and changed into fresh drinkable water by the staff of Moses.*

- **Exodus 7:22-** *And the magicians of Egypt did so [turned <u>water into blood</u>] with their enchantments:*

- **Exodus 8:6-** *And Aaron stretched out his hand over the waters of Egypt; and the <u>frogs came up</u>, and covered the land of Egypt.*

- **Exodus 8:7-** *And the [Egyptian] magicians did so with their enchantments, and <u>brought up frogs</u> upon the land of Egypt.*

- **Josephus-** *Moses put the rod down upon the ground, and commanded it to <u>turn itself into a serpent</u>. It obeyed him, and went all round, and devoured the rods of the Egyptians, which seemed to be dragons, until it had consumed them all. It then returned to its own form.*

131

Below is a table of the "known" magical things accomplished while holding a staff. Moses was the king of the magical staff. This does not include general forms of levitation called out in many texts because specific people were not identified.

	Moses [Jew]	Mises [Sumerian]	Malka& Musa[Arab]	Elijah & Elisha [Jew]	Jajamankh-[Egypt]	Bimater[Babylon]	Thoth [Atlantean]	Jannes [Egypt/Jew]
Raise Water Of A River	■		■	■		■		
Water from Rock	■		■			■		
Turn bad water to good	■			■				
Turn Rod To Serpent	■						■	■
Call Frogs	■							
Turn water to blood	■							■
Used As Weapon	■						■	
Call Locust, make it dark	■							
Heal Serpent bite	■							

132

Apes, Egypt and Electricity

What do Apes, Egypt, and Electricity have in common? The answer or at least the question is found on pictures in a temple in Dendera as shown below. These pictures depict something people like to call the Dendera tube. Two of these devices are shown below.

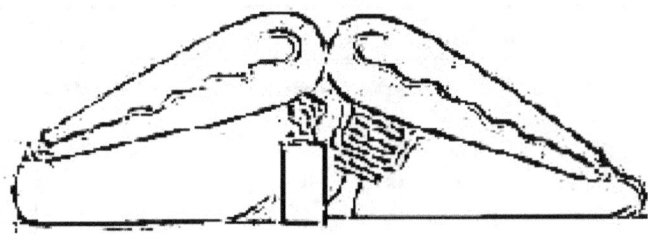

The devices appear to be electronic and apes typically are depicted with them.

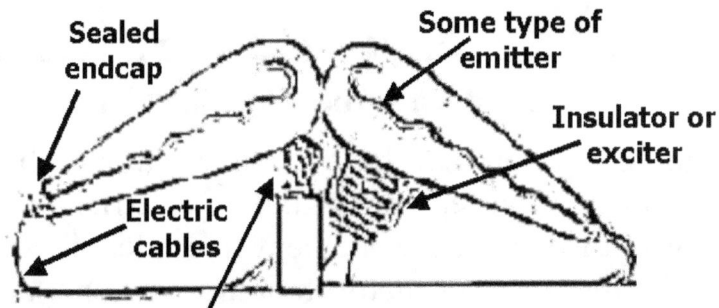

Sealed endcap

Some type of emitter

Insulator or exciter

Electric cables

Tiny little person getting close to the emissions

If that were the only depiction, no one would think too much about it, but more and more were found. The following group of pictures below depicts some type of electronic tube-like objects resting on still other strange devices. A little human who tries to point the main device at a baboon holds all these up. That's right; a baboon. The first shows a portion of the wall etching while the drawing to the right shows the full scene including the baboon holding up knives to protect the Dendera.

I'm not getting into the whole baboon thing in this book, but I will tell you that the ancient Jewish book of Jasher indicated that when the Tower of Babel fell, 1/3 of the people were turned into baboon like people and the Egyptians were chipping away stone to depict baboon-like people. OK; forget the Baboon and concentrate on the electronic tube thing.

I said forget the baboon. That is not part of this book.

Dendera Tube and Electricity

I've got to tell you that these "Dendera tubes", by their very looks, must have used some kind of electricity. The twisted cabling to their base, the filament like internal structure, and the radiation type insulator holding the one on the right below all point to the same conclusion. If you recall I presented a case for the Great Pyramid producing electricity for Egypt and other places. This could have been one of the devices that used the valuable resource. Cables are attached to the "Dendera tubes".

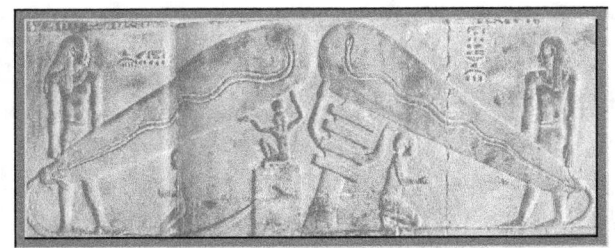

The electricity would have been used to make these things radiate in some way. The radiating component will be explained later, but we should, at least, recognize that the tubes produced something special.

There is something else you should recognize. There is a miniaturized image of Pharaoh under the left Dendera tube on both of the first two carvings. This indicates that the pharaoh needed whatever this machine produced and the pharaoh was insignificant with respect to the Dendera Tube. In the third image below is only different in that the Pharaoh is under the one on the right.

These things must have been really something to be more important that a pharaoh.

The first and third images show a huge baboon guard wielding a knife, evidently to protect the Dendera tubes, so here is my theory, for what it's worth.

Maybe the thing had something to do with keeping the demigod rulers [like pharaoh] alive after the tree of life was lost in the worldwide flood. The Baboons were hired to make sure most people did not live as long as the rulers.

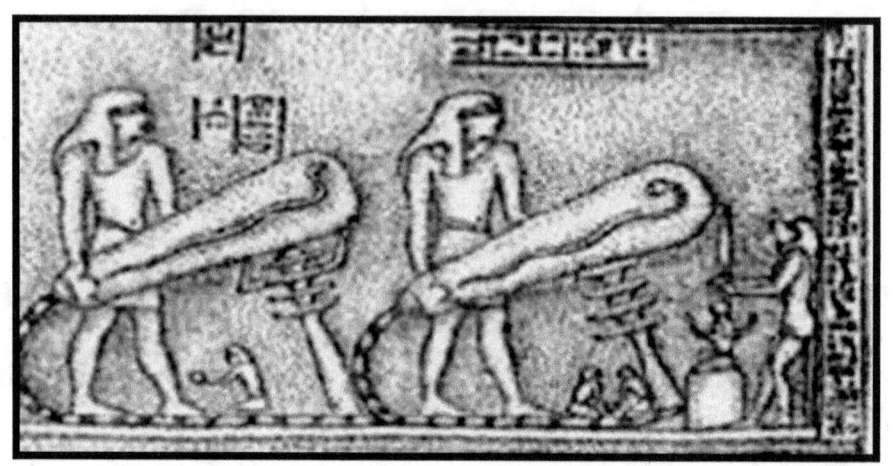

The way one could affect life with some type of radiation is by affecting the vibrational element of life itself.

I know you think that DNA structure determines the life cycle, but scientists today know that DNA has no control over such things. Nor do light bulbs, but these were different than the normal Egyptian lightbulbs shown below.

Surely, the addition of human growth hormone has been shown to reduce the effects of age, but the method for its reduction and the reasons why some children die of old age when they are still children has never been explained by DNA. We will get to that later because we first need to understand many were using these Dendera radiators.

More Energy Devices

Sumerian Radiators

Not only do we find these things in Egypt, we find them around the world. The picture below appears to be a similar type of thing and it is being reverenced by 2 regular guys and 2 with wings. I know the wings are curious, but the thing to look at here is the flying ship carrying a person just above the Dendera tube. Versions of this same picture started showing up all over the Middle East during very ancient times. While the tube must have had something to do with flying, there seems to be a strong desire for the "people" in the drawings to be close to this thing as well. Someone concluded that the thing in the middle may have been responsible for keeping the "gods" young.

Read the section below from the "Epic of Gilgamesh" and see if this could have been describing the Dendera Tube energy. They are always revered and protected as they were depicted around the world. After all, these "tubes" may have been able to vibrationally change life itself or some equally important function.

Sumerian Epic of Gilgamesh- *Upon her corpse that was hung from the pole, they **directed the pulse and the radiance**; 60 times the water of life, 60 times the food of life, they sprinkled upon it; and Inanna arose **[Sounds like some kind of electric shock treatment, used to bring Inanna back to life. Could the pictured apparatus have been the device used to deliver the directed pulse??]***

In this case, the individual wanting to extend his life got inside the radiator. The first image shows Inanna inside what they called their Sky-chamber that brought her back to life.

Another Sumerian Radiator

The depiction below is very similar. This time it is men dressed up like fish that are trying to get close to the radiator. On this seal can be seen a tube with some kind of rays emanating away from it. Like others, the device looks very much like the Egyptian Dendera Tubes. We can call this one the "Sumerian Tube". The only difference is that fish men have replaced the baboon and one of Sumerian Merkabas [Aircraft] is usually flying overhead. For those not knowing about Merkaba, they were flying transport vehicles described all over the place in the Middle East. I would assume that they were powered by cinnabar and serpent slough like the Indian Vimana flying machines described in "*Mahabharata*".

Still Another Picture of the Device

Another picture of the ancient Sumerian form of the Dendera device is shown below. The last one, right, is an Egyptian image showing they also had similar flying machines.

Really good Dendera tubes might have been used to help the flying machines rise off the ground by some type of massive levitation thing. The fish suits have been replaced, but the depictions are always the same. If you look closely you can see men flying in the things above the Sumerian Tubes. Possibly, Dendera tubes caused levitation by changing the characteristics of matter and life.

Assyrian Radiating Tubes

Here are some examples of the Assyrian version. In this depiction, the radiator seems to have fruit making it more probable that the Nephilim made artificial "Tree-of-Life fruit with the device. A Merkaba [Flying Machine] is still dancing overhead just like in the Sumerian version.

Assyria Fancy Model

Here's an interesting one. This time it's from Assyria. The men have changed clothes, but the rest is the same. The flying machine has 3 men riding inside. The Dendera tube looks a little modern and one of the men has encased himself in one of the tubes while a second one under the flying machine looks like it has little lightbulbs on it.

Assyria Again

In the next Assyrian artifact shown on the next page, the Dendera tube has sprouted flowers and Cherubimic eagle-headed "angels" are picking the fruit. This could have been showing that the Dendera tube was an artificial Tree-of-Life as the eagle gods pick something that looks like fruit from the radiator and there is certainly a reverence to this flowering Dendera tube.

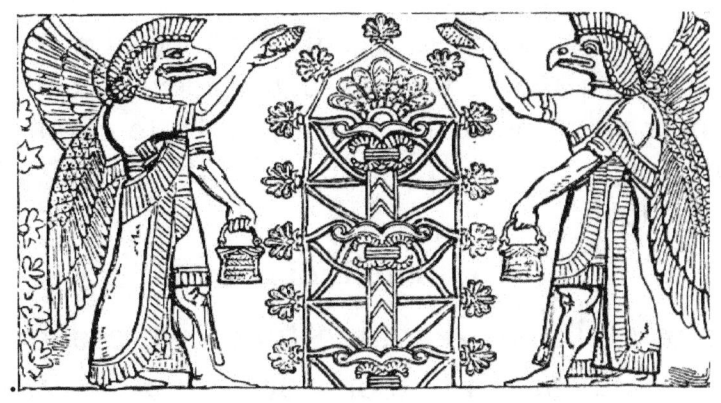

Central American Radiators

The Aztecs and Mayans both depicted "radiating tubes" of some kind. Something shoots out of the Aztec version while the Mayan one shows an internal filament similar to the Dendera tubes of Egypt.

Aztec Dendera Tube

The Aztecs must also have known about this device as well. The Codex Nuttal shows the device. [See below left] Some kind of radiating beam is coming from the central orb. The device is standing by what appears to be a rocket and a throne. No people are in this section of the work, nor are there any baboons, but a baboon was one of the Aztec gods.

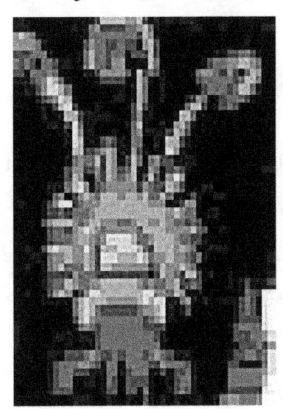

Guatemala Dendera Tube

The Mayans also depicted a radiating tube of some kind as shown in this page of the Dresden Codex. [Above right] I don't know what the snake thing is in the foreground either. Two filaments can be seen along with the radiating "Glow" around the tube.

Indian Radiator

In India, a similar radiating tubes were depicted, as the type that had a person get inside the tube. [See below left and center] It certainly was no simple light bulb. [Did you remember the Sumerian guy that also got inside some type of chamber when he was near the Dendera tube thing?]

Uzbekistan Dendera Tube Inverted

Just like the one found in India, Uzbekistan was the location of the rock carving shown on the drawing above right. Can you see the similarity? Is there someone inside some radiating chamber? I know there are a couple of odd looking guys holding on to the thing, but there is a person inside.

Egyptian Modified Dendera Tube

Almost exactly like the Sumerian version, the Egyptian model changed as shown below. Notice the identical flying

142

thing next the radiator and the complete reverence of the gods. I don't see any fruit in this one, but there can be no doubt of its similarity.

Later we will see a new symbol that came out of these modified Dendera tubes. The new model will be called the Djed. Right now, I only want you to see the striking similarity of these emitters and begin to understand that the ancient people knew how to manipulate things with electricity and quite probably they could manipulate matter by high-energy vibration of this electrical field.

More Vibrational Items

Before we get back to the details of vibrational "everything", let's talk about a few more seemingly magical items called a magic stone, the Ancke, and the Djed. I know they are crazy names, but they must have been powerful tools just like the Dendera tube. These items have been written about in hieroglyphics for centuries, but no one really knows what they did. Along with the Uas [staff] these other strange devices must have been the catalyst of many seemingly magical capabilities. Later we will see how there is a reemergence of these unusual items or at least we might begin to see what these tools might have done for the ancient "mystics".

As I indicated earlier, the manufacture of electricity and worldwide distribution of electricity, most likely, was still in existence for about a thousand years after the Tower of Babel was destroyed sometime around 4000 BC. Finally the Pyramid quit working, but it must have been instrumental and crucial in getting many of the "Magical Devices" to work. Today, all we are left with are symbols of these marvelous items.

Magical Stone

The ancient Jewish holy book known as the Talmud says, "God hung a precious stone around the neck of Abraham, all that were sick and gazed upon it were healed. After Abraham died, God took the stone." That doesn't mean that getting a crystal necklace with keep you well, but the reference could be talking about an actual device. Who knows? And if that reference isn't enough, let's look a few more.

Greek Evidence

Lucian- This Greek Satirist from around 150AD wrote that in Hierapolis, Syria he saw a shining jewel that illuminated the temple at night and that glowing "Stones" kept light shining all night at Baalbek.

Later European Evidence

Saint Augustine- He wrote about a wonder crystal lamp in Egypt. He said that neither wind nor rain could extinguish it. [~400AD]

Hopi Indian Evidence

Hopi Indians- They used a Light generator made from a rectangular piece of luminescent quartz and a bolster shaped upper piece worked by rapidly rubbing the crystal, according to researcher David Chandler. These lighting "machines" have been found in several locations in southwest America. No, they haven't found a working one yet.

The Djed

It seems that the Dendera tubes shape slowly was modified or depicted differently. Still reverenced by winged "gods", the outer shell was all but gone and the radiating fins took on their own shape. This new form of radiator was called the Djed or [emitter of stability]. The Egyptian Djed looked like a post with air-fins and was the **second most common magical implement** of ancient times and widely worshiped as can be shown in the following group of pictures and drawings. According to the pictures it was evident that even the angels worshiped the Djed. We may never know what was so important and significant about the Djed. The Djed, apparently, was revered as a levitation device and a life-giving device, especially when used in conjunction with other devices to be discussed. I already showed one of these things during the discussion of the Dendera tubes. Possibly it was a newer form of the Dendera tube or some similar item. It appeared to be more portable in many depictions.

Not a Backbone

Modern day Egyptologists simply say the Djed was a symbol of a backbone. It does kind of look like a backbone, but beyond that the description seems absurd to me. Paintings of the device have been found on more walls of Egypt than any other magical device. Only the [UAS] staff is shown more abundantly. Literally thousands of drawings show this device. Note how the winged angels worshiped this strange and magical component. It is my opinion that they are not worshiping a backbone but something that was very special during ancient times. Even though we may never find out what it did, there is no doubt that it was of huge importance to the ancients in Egypt.

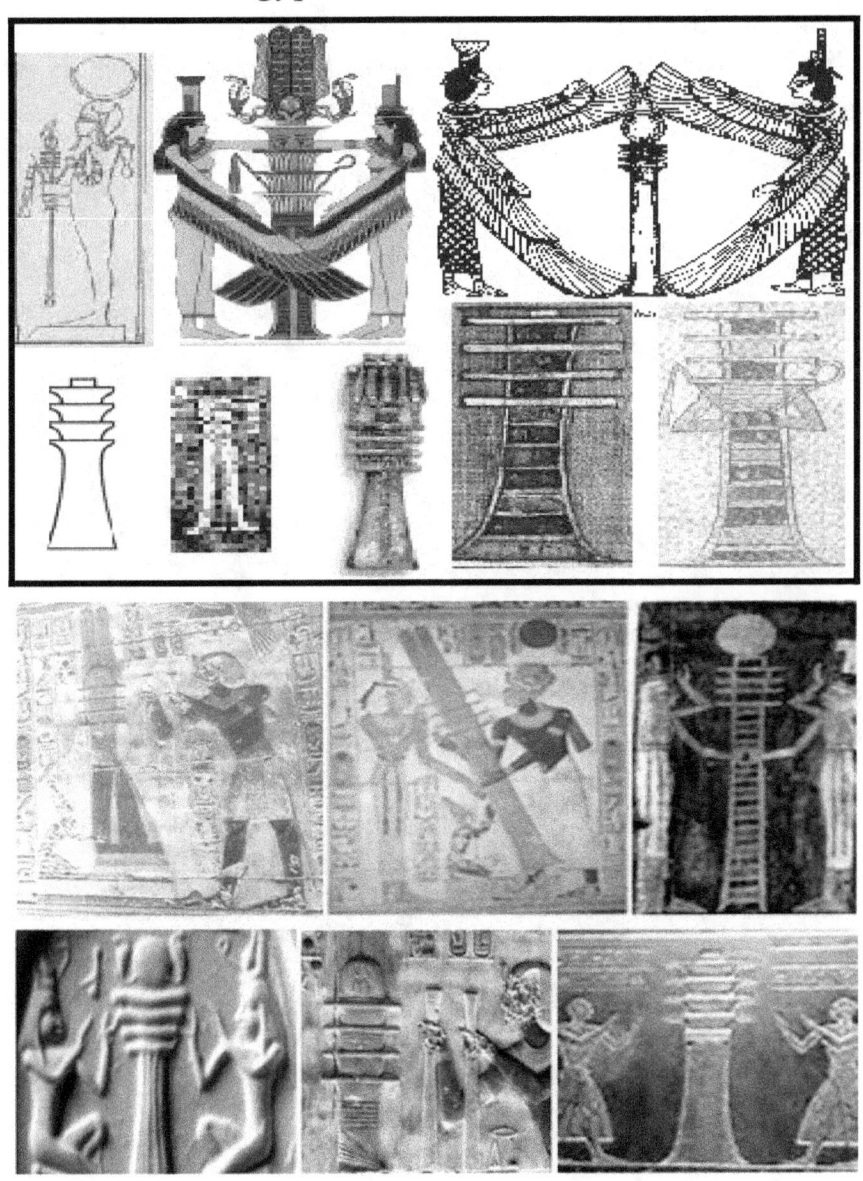

> *The Djed was revered my all the gods because it could do something for them. The level of reverence of the gods to this thing tells me it was part of the life extending machines or plants.*

If you notice in the next picture, not only are the rulers worshiping this Djed that appears to have an "ancke" on top holding an orb that may simply be a Dendera tube as seen from the top. Please notice 6 apes are worshipping it as well. I'm not going to go through the descriptions of what this might mean, but please recognize that anyone, and anything, that was important worshipped the Djed.

Sumerian Djed

Don't believe for a moment that Egyptians just made up this unusual artifact. In ancient Sumeria, we can find the same item drawn and etched on walls. In the next group are some of the many depictions. Some had three spires while others had 4 or 5, but there can be little doubt that the devices are the same as those depicted in Egypt. On occasion, the Djed-like device was depicted with flames coming out of it top as shown to the left. Apparently, a "magic staff" may have been required for the flame.

The depiction is further pronounced in the following sculpting of the Sumerian version of the "king/sphinx". Notice the "lion ruler" is pampering what appears to be a garden containing plants that look like the Egyptian Djed. Possibly, it isn't full grown, but it seems to be very important. I don't think Sphinxes cared about backbones but they certainly cared about djeds.

Levitation and the Djed

Here's where the DJED is connected with vibrational levitation or modification of elements. Back in Egypt, we find some more references to Djed. The picture-word "Djed" is generally depicted as the symbol "medu", which is usually interpreted as making the single sound "SAH". From the text below we can assume that the Djed-device may have also had something to do with making a special sound that

149

allowed for levitation, just like that described in earlier examples of levitation. Actually, it is not a difficult leap to believe the DJED had something to do with Flying. These excerpts below come from the Pyramid Text "utterance 539. Let's read the verses to see what I mean.

- *[1303b]* <u>***A Djed making the SAH sound***</u> *will allow him to levitate and become empty on the way to the Blue Void.* *[**Somehow the Djed and levitation are linked.**]*

- *[1304c] (A Djed making the SAH sound), will allow the sight of Pepi I to be as "the Opener of the paths"* *[**The Djed also seems to produce radiation of some kind.**]*

- *[1309a] (A Djed making the SAH sound), will insure that the two shoulders of Pepi I will be as Illusion.* *[**The Djed seems to be associated with invisibility.**]*

- *[1313a] (A Djed making the SAH sound), will insure that the buttocks of Pepi I will be as the Right-levitator together with the Left-levitator.* *[**Somehow the person can steer with Pepi's butt? I'm not going to explain this one. Just watch where you sit when you are at Egyptian ruins--- especially if your butt cheeks are not coordinated.**]*

- *[1315a] (A Djed making the SAH sound), will insure that the soles of Pepi I's feet will be as two correctly-positioned levitators.* *[**Possibly it is not talking about the person Pepi, but some kind of ship with flames coming out of its base which would be similar to the feet. Then the butt makes more sense.**]*

- *[1316a] (A Djed making the SAH sound), will insure that Pepi I will be one who is on the way to be a god, or a son of a god.* *[**Seems to be referring to going into space.**]*

- *[1317a] (Whenever the Djed is properly used), Pepi I will be reincarnated for the god-Star. **[Again space or heaven are referenced.]***

- *[1318c] (A Djed making the SAH sound), will insure that Magic will be in the body of Pepi I. **[Assuming Pepi was a person, the Djed had something to do with producing Magic.]***

- *[1320c] (A Djed making the SAH sound), will insure that there will be Second Sight and the Youthfulness of a young person. **[This seems to indicate that the Djed somehow controlled aging.]***

- *[1324] It is not Pepi I who speaks to the gods; it is Magic, which speaks to the gods. (A Djed making the SAH sound) will insure that Pepi I reaches the lower realm of Magic. **[The Djed was strongly connected to Magic.]***

- *[1325c] (Whenever the Djed) is heard, the gods will provide him a seat in the levitating-god-Star. **[The Djed somehow made the user able to ride in the "levitating god-star" I have no idea if the seat has butt controls so just get that image out of your mind.]***

- *[1327a] The gods will accept the control of Pepi I on the way to the Blue Void, because he has already gone to the house of Second Sight, which is in the Calm Void Current, so that the nature of his consciousness will be true to the Soul. **[Even control of (the spaceship?) was possible.]***

The Djed Again

Utterance 239, talks about Djed medu just like the preceding one. Here is what the texts had to say.

- *[243a.] A Djed produces the sound "sah", so that the White Crown Crossing Power will engage in Levitation. It is because it has become aware of the Mahatmic Eye.*

[The Mahatmic Eye is associated with second sight while levitation is clearly raising an object off the ground. Both seem to be associated with the Djed. Other than that, I don't know what the verse is trying to say.]

I don't know exactly what this thing was, but it was something amazing and don't let people simply tell you it was nothing more than a symbol of Osiris's backbone. That is absurd. From the above verses, we can see that in addition to the possible connection with extending life, the Djed was thought to have many other capabilities including levitation, magic, second sight and controlling someone's butt. No wonder it was so revered.

Tuning Fork Ancke

The Ancke was still another magical vibrational item revered by the ancient Egyptians. It looked like a cross with a loop at its top. It was just about as important as the Djed. The present-day Egyptologists tell us that it had something to do with eternal life itself. Again, no one knows what it was for sure. The pictures of Anckes are everywhere almost to the level of the Djed thing. Its popularity must say something about the power that it once yielded. The early Nephilim and demigods that survived the flood kept them close and pictures of the gods almost always show this magic device with them. I don't want to speculate further about this device, but don't discount the fact that this device must have been very useful at one time. It was not simply a pretty symbol that was lugged around by the ancients, as some would have you believe. Below are some of the hundreds of depictions.

Ancke and Uas Together

There seems to be something even more special that happens whenever an Ancke and Uas were used together as shown in the following group of pictures. From the pictures, there

153

seems to be a strong connection with the two items being used together and the possibility of bringing people back to life. The picture in the upper right corner is of Thoth, the creator of the pyramids, the great teacher, and the first of the Egyptian magicians. He was certainly a demigod and he is using great care when carrying the Ancke and Uas in combination and always had one or the other with him. Note, the bottom picture shows a dead pharaoh lying on a bed of these objects to apparently aid him in his resurrection.

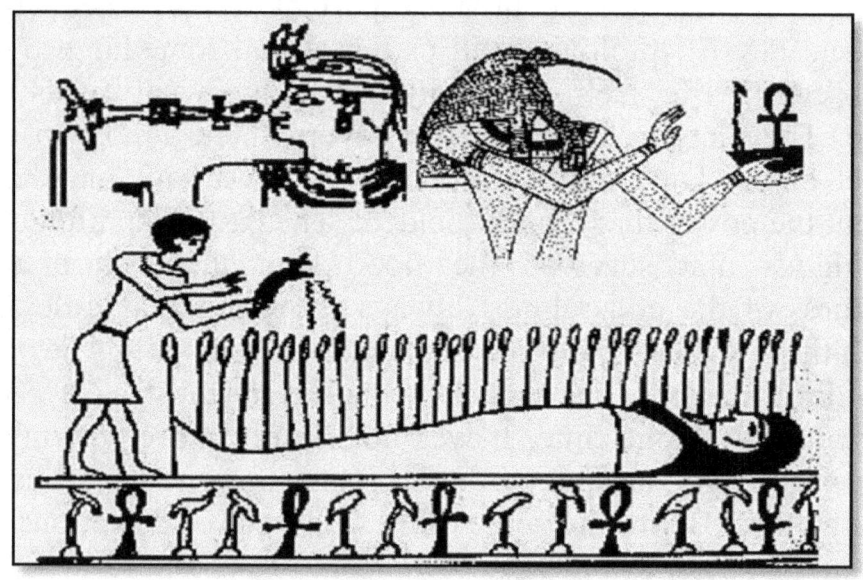

Ancke, Uas and Djed

If two of the magical items were powerful, just think what all three must have been able to do. Many pictures show the Uas, Ancke, and Djed all working together. They seem to be some kind of weapon in this configuration. Evidently, these items had a substantial amount of power when originally wielded by the ancient rulers. Over time, the shape of the item was worshiped because of what the actual item had been capable of doing in the past.

Just as a note of caution, don't mess with anyone carrying around all three of these objects.

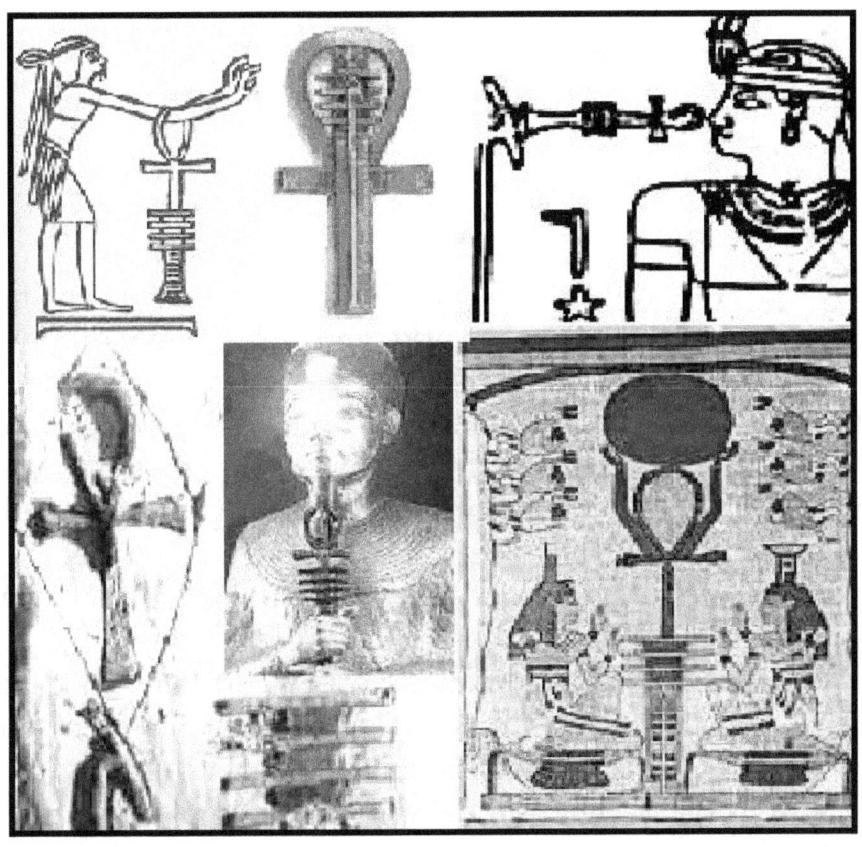

Arc Power Anomaly

As we travel away from Egypt, we find more clues of both electricity in the ancient world and some type of power associated with vibration. There is strong evidence to suggest that the Israelites used the Egyptian made electricity to help power the "Arc-of-the-Covenant", and other "magical devices". Certainly, the arc's shooting out lightning, reported in ancient texts, seems to be a sign of a buildup of electricity inside a large storage device. The major question, in the past, has been, "How did the electricity get there in the first place?" The ground current electricity theory may shed light on that anomaly. Other feats of "what was reported to be magic" probably also needed some ancient form of electricity. Even the use of the "magical staffs", during ancient times may have needed electricity that probably was provided this same way. It was like having portable tools without needing batteries.

Moses

Let me back up just a little so that you can understand all about how Moses made this Arc thing and about his "magic" staff. Certainly, we can say God gave him the power, but that may not be the whole story. If we go back to the time of Joseph in Egyptian history, he was governor during the 13th Dynasty. It is believed his pyramid is in Sakara and his name was Kendjer but that is another story. Soon after Joseph and possibly his brother Benjamin had been governors, The

Hyksos [Amalekites] invaded and took control of Egypt. These people came from the land of Canaan, but they were not from the clan of Abraham. For a long time, the Hyksos [15th Dynasty] and Egyptians [14th, 16th, and 17th dynasties] sort of ruled together. The "power plant" pyramid was already thousands of years old by the time the Hyksos came along and I would suppose that the earth's resonating frequency had changed drastically. Electric equipment could not work as efficiently if they worked at all and other hardships may have been creeping up, but evidence suggests that there was still a lot of electricity being generated even during the time of the Hyksos and during the great wanderings of Moses and the "Israelites". As the economy started to decrease, unrest lead to the enslavement of the Jews for a time and the eventual downfall of the Hyksos as Moses killed the last of the Hyksos Kings [Apepi] in the Red Sea. Apparently, Jewish Magicians still held onto a high level of status including two Jewish magicians named Jannes and Jambres during the reign of Apepi. We can believe the entire order of magicians was "controlled" by the remaining Hyksos magicians. Moses. most likely. became one of these esteemed magicians as his princely position during the reign of Khyan. The "magician's sect" was a mysterious lot. These guys experimented with VIBRATION and Electricity. There are several reasons why we know about the vibration stuff, because all three of these magicians [Moses, Jannes, and Jambres] made sticks turn to snakes, turned water bloody, and called frogs to infest. If these things seem odd and impossible, we can either believe Biblical history or not. If we believe it, Moses, Jannes, and Jambres and other contemporaries, could convert matter into other forms of matter. The way John Hutchison is doing this today is by using vibrations. We can assume they used the same form of MAGIC during the time of the Pharaohs.

157

Moses Leaves

When Moses and his team left, he did not forget all he was taught about using the electricity from the Pyramid nor did he forget how to change Matter by subjecting it to various vibrational elements. After Moses constructed the magical Arc, I would assume that it was "charged" by the output energy from the pyramid and the "focusing "staff" he carried with him always. Possibly, he simply modified the materials inside the arc to produce ultra-high voltage Electrical charges by means of focusing various vibrational electromagnetic waves from his staff. At the <u>right combination of vibrations</u>, the "essence" of the Arc became a weapon and it spewed destruction. I know this isn't the story you heard previously, but the evidence suggests this is a more responsible discussion and it shows that vibrational modification of particles was well known during that time and the electricity required to initiate a vibration could, very well, have come from the thing we call the "Great Pyramid".

This small device, covered in gold [the great electrical conductor], not only was said to have housed the commandments of God and other sacred items, but it also killed many people, according to many references. In order to get close to it, people had to remove their shoes. After a while it got so dangerous that the Jewish people didn't even want it near, even though it had been responsible for several of their victories during battles against their enemies and it had made it possible for the Jews to cross the Jordan River. Along the top of the arc were golden cherubim, but it was certainly more than a box. After all, it was designed and made by one of the highest-ranking members of the royal family in Egypt and presumably, he had been schooled in the arts of magic just like many of the other royal personnel. One of the "Magical" capabilities was, most likely, the ability to produce electricity.

God Made Fire

If you don't accept the electricity theory, there could be other explanations for the Arc's miraculous power, but just saying, "God focused energy through it", is probably not the way God has worked in the past nor is it probably the way he helped the Jews with this device. His grand order includes physical laws that he works through. We've seen it millions of times every day. Possibly, the only two times he did some type of creation that was separated from known natural physics was at the creation of Adam and the embodiment of Jesus, and those occurrences probably were governed by some natural law as well. We just don't know what it was at this particular time.

In addition to sending out electric charges, the device could cause rivers to dry up and kill anyone closer than 2000 cubits **[about half a mile]**. There were indications that lightning came out of it and death came to those who touched it. It got so bad, that whenever others captured the arc, it was quickly returned to the Jewish people, and the Jews didn't even want it near them because they feared its power. This thing was like a huge electrical capacitor, which possibly stored thousands of volts of electricity.

- *Joshua 3:3-4- And they commanded the people, saying, when ye see the ark of the covenant, and the priests bearing it, then ye shall remove from your place, and go after it. Yet there shall be a space between you and it, about **two thousand cubits** by measure: come not near unto it, that ye may know the way by which ye must go*

- *Joshua 3:15- And as they that bare the ark were come unto Jordan, and the feet of the priests that bare the ark*

159

were dipped in the brim of the water, the waters which came down from above stood and rose up upon an heap **[The arc also moved water away just like Moses' rod had done earlier.]**

Central American Electricity

On the other side of the world they were getting electricity from somewhere and they also were able to levitate things and other seemingly miraculous things because these people apparently knew about how to manipulate matter with vibration. For this overview, we first travel down to Mexico City and there we find ruins of the Aztec and PreAztec. It is easy to see the difference in construction of the Aztec structures and those built by very ancient people living here well before the Aztec. It is believed that the preAztec people lived in the Mexico City region for as much as 10 thousand years or more. These guys had science and capabilities we have not yet re-learned. Two such buildings can be found here. The image following shows very strange parts of the city ruins at Tenochtitlan.

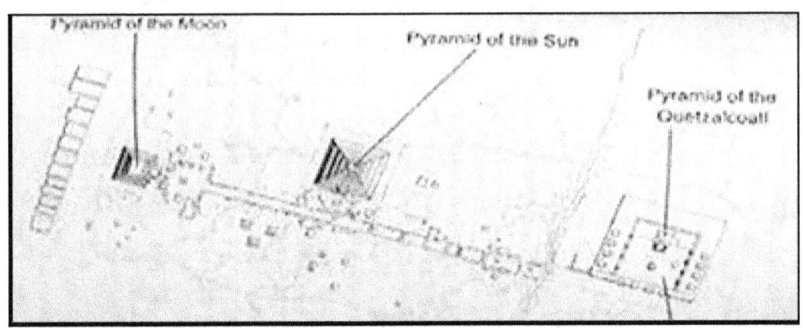

One is called the Pyramid to the moon and the other is called the Pyramid to the sun. Between the two is an underground corridor that was an ancient lava bubble. Now it is simply a tunnel, but inside the tunnel was found sheets of mica. Let me tell you what Mica is for. Mica is used for electrical insulation. It is too brittle for much else. To make this odder, the huge mica sheets were evidently transported thousands of miles as they were quarried in Brazil. The previous drawing shows the two preAztec buildings and the underground causeway, but there is more. Around the Pyramid of the Sun [in the middle], the mica sheets were continued all the way around the pyramid itself. As we continue to investigate, we find MICA sheets in the pyramid as well. The mica sheets were placed in between floors in the massive Pyramid to the Sun. Between the floors are also unusual conduits as described by the graphic following.

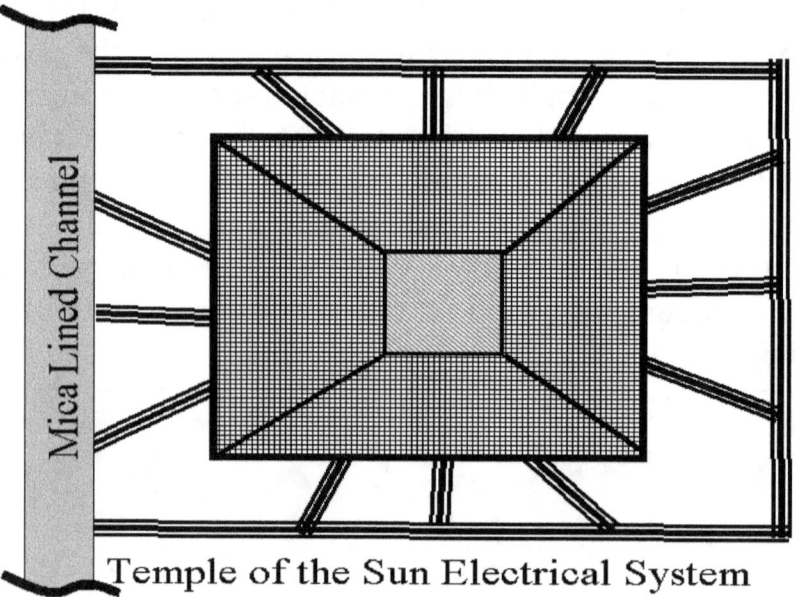

Temple of the Sun Electrical System

Electrical Conduits

Conduits have been found throughout the building to the outside. Initially the ducting seemed to be for water transport, but sometimes the ductwork branches in more than one direction and at right angle as though large cables had once been channeled from the electricity building. Next, we see images of a very small group of the strange conduit units put together as they might have been and a through-the-wall element which would allow the removal of electricity or entry of cabling, who knows?

Adding up the indicators makes the probability of electricity storage and use a reasonable suggestion. What electricity might have been actually used for is somewhat of a mystery, however, there is evidence that it was used for electroplating metals together, for lighting the ancient buildings, and for levitation. I'm sure there were other things, but it's very hard to find physical evidence of things that happened well over 5 thousand years ago as just about everything disintegrates over time.

American Neat Buildings

I'm not getting into the impossible interfaces of huge boulders that are found on the building made by those who came before the Inca and the Maya. Everyone tells you about them and it is easy to see how levitation would almost be a requirement for their manufacture. What I want to talk about here are just a few things that show strangeness. The first is a circular building. Near the huge triple walls of

Sacsayhauman, in Peru, we find something very interesting and still undefined.

The preceding picture shows what the walls protected. This is a perfectly round building-like structure with about 21 rooms in an outer ring, 11 rooms in an inner circle and a huge, perfectly round central arena. The problem is that there are no doors in any of the rooms. There is no doubt that this was an extremely important object of the pre-Inca, but to say, as many have, that it is this huge astronomical calculator, sounds like hogwash. To the right I am showing the remains of a nuclear plant in Russia. Please notice the similarities. The following images are of the seemingly impossible to build blocks on one of the three sets of walls that can be seen at the top of the image to the right.

No one has explained how this huge complex could be used, why it was built, and why it had to be so well protected. The picture below shows another view of this mystery. This may or may not have much to do with vibrational matter, but I thought you would like to see it so that you could begin to make judgments about what it might have been and why it was so important that it lay behind three sets of Herculean walls as shown in the aerial picture following. By the way the blocks look like they are grown in place and some type of vibrational manipulation might be the answer to that mystery as well.

Speaking of strangeness, if we move a little south near the Nazca plains [where the hundreds of images are found on the ground], the carved 180-meter-long thing shown next can be found, carved into the side of a large hill. Some say it points the way to the Nazca images, but certainly, it has no pointing element. Most call it the Candelabra because of its general shape, but that isn't it either. Whatever it was; it was so revered or was so important to the ancient Americans that they spent an enormous amount of time producing its image. Strangely. Like the Nazca lines, it can be seen much better from the Air. I know you are thinking it looks a lot like the Djed thing so I also have an almost identical device etched into stone by the Sumerians. [see right]

American Levitation

While some of the enigmas of these ancient people can't be easily answered, possibly one enigmatic thing can. The thing I'm talking about here is the fact that no wheels were used to support carts and vehicles, according to just about every investigator. The reason no wheels were used is that no wheels were needed. Instead, they most likely used some form of levitation. I know this probably sounds absurd and just not having wheels doesn't mean people could levitate. So, you might wonder why I even suggested such a thing. After all, one would have to manipulate matter to allow heavy materials to float in the air.

We now <u>can be fairly certain</u> that, during this time, levitation was commonly known as no wheeled service equipment remains have ever been recovered from this era, while wheels were used on multitudes of toys. Below are some of the many found as wheels were well known and almost never used.

Haven't you ever wondered why the early Americans never used a wheel?

Well, the most logical reason is that they didn't need to.

Even after the Babel incident, no wheels were needed, because they were re-taught levitation. If they weren't, how can we explain the wheeled toys that have been found in an area completely devoid of wheeled carts. Here are still more of the "wheeled toys" from the Central and South American areas.

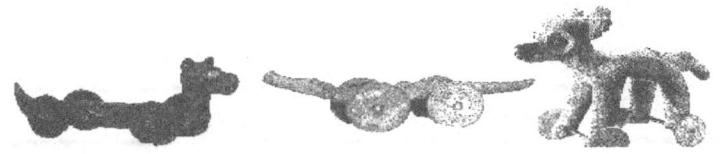

I know 5 stupid wheeled toys show people knew about wheeled locomotion, but they started finding more and more and more. Another 16 are shown next.

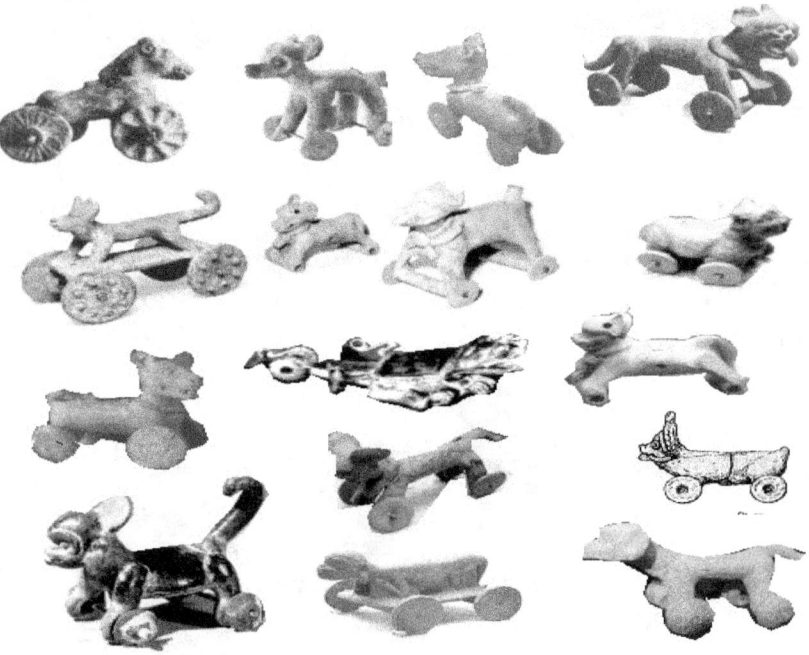

Incan Flying

Here is some more evidence that the Pre-Inca didn't need wheels because they had another way to get around. Pictorial evidence was found at the grave of the Lord of Sipan in Peru. The drawing below was taken from the site and seems to show some type of flying craft with flames coming out of the rear. Following the craft are three smaller craft carrying people. Please notice there are no wheels on any of the vehicles.

Who needed the wheel!!!!

When we add in the rider of the front vehicle, it can be easily seen that this is depicting a self-propelled vehicle with no wheels. It either glided or flew, but wheels were not required in ancient Peru.

Broken Pyramid

After a period of time the Ancke, Djed, Uas, and Dendera tubes became emblems rather than useful items, as people forgot about electricity and forgot how to manipulate matter. We know that the Pyramid power plant slowly became nothing more than a strange building by 3500 years ago. A possible reason for this reduction in electricity might be found in the earth itself.

The Broken Arc

We know, from reading the Biblical record and simple observation, that the "Arc" slowly lost its power over time, but it was not until over 500 years after the time of Moses. No one can really tell you why the arc, broke, but people tried to come close to it one day about 800 BC, and they were not killed. Soon, its magical importance was gone. Apparently, the Arc was not the only "vibration focusing or amplification" device that was around during the time between 5000BC and 1000BC. All types of odd devices apparently captured energy and initiated changed to particles we have only now begun to understand. We will find out that people used the electricity or the Great Pyramid to power a wide assortment of almost magical devices like magic staffs, Anckes, Uas, and the famous Dendera tubes. By 3000 years ago or so, everything came to a halt. The Pyramid quit working and even before it quite delivering electricity, all knowledge of how to manipulate matter by changing its vibrational component, had been forgotten. Why people forgot about the once great capabilities of the past and how matter could be manipulated is another story that involved and ancient worldwide struggle called the Bharata War. It was said that 1/3 of all the people of the world died during this horrible war and something happened during the hostilities. Human DNA was mutated and Haplotype scientists have traced the date of the massive change to about 5100 years ago near the end of this war. The massive mutations started a decline in man's capability and memory and brain size. [Cro-Magnon, Neanderthal and Homo-Capensis human brains were all larger than our present "Modern" brains.] After the War, people still had electricity from the Pyramid, but when the pyramid lost its power, there was no one to regenerate the electricity to rejuvenate and rekindle the useful tool. They didn't recognize that the earth itself had changed.

Earth Vibration Change

The Earth vibration frequency has been changing over the years. Scientists used to think that its frequency stayed at 7.8 Hertz, but new measurements has shown that its base frequency has been increasing since 1980, for some unknown reason, to about 12 Hertz. Maybe, the earth's base frequency changed, which kept the transfer of electricity from being as efficient as it was when the pyramid was originally built. With electricity harder to make, the Egyptians got tired of pouring in the liquids necessary to keep it running.

Changed Resonance

Another reason for disuse may have been that the pyramid finally worn-out. Researchers have found one of the granite crystals broken and the size of the oscillating chamber has had the walls move out slightly over time, which would change its resonance. Some of these factors and others, we have not recognized, made the pyramid less and less efficient, until it probably just quit working on its own. Don't discount this seemingly fanciful theory. There is real evidence that electricity was used during these ancient times and it had to come from somewhere--. Why not the Great Pyramid? With electricity, scientist could cause electromagnetic vibrations that could affect, forces, Matter, and Life itself as all of these things are vibrationally based.

Levitation Today

Maybe some of the knowledge about vibrational control of matter still survived around the world. Today and in our recent history we had no Great Pyramid to provide electricity, but the knowledge of vibrating crystals and levitation was not completely lost over time. Generally, levitation is discounted as an absurdity or trick because normal 'non-vibrational" theories can't explain this type of action. Things can't suddenly lose their weight. Weight is dictated by mass in a specific environment and mass can't be changed. The environment also cannot be changed or the object would not be located where it was before it levitated. Possibly these anal-retentive researchers are limiting their viewpoint, the Tamashii model and a unified particle will come to the rescue.

Levitating People

This next section doesn't seem possible, but the evidence is well documented and witnessed, in many cases, by large groups. Many people have had the ability to lift their own bodies into the air over the years as indicated below. No, I'm not talking about David Copperfield here. Some of them lost their lives because it was considered to be an act of the devil, but some survived and floated around. If this list seems impressive, just think of the numbers we would know about if levitation wasn't considered demonic in the old days. Arthur Herbert Thurston wrote the "Physical Phenomena of Mysticism" about many instances of levitation. Some of

172

those listed below came from his investigations and other sources are otherwise identified.

St. Bernardino Realino- *He initially had a glow come from his body before he rose in the air in 1608 according to testimony of a nobleman of the time.*

St. Adolphus Ligori- *He was supposedly raised into the air in front of his entire congregation in 1777 according to the same work.*

St. Joseph- *He would cry out and be lifted up without regard into trees and above his congregation, according to Catholic traditions.*

Father Francis Suarez- *He reportedly began to glow with a blinding light before being lifted, according to a catholic monk.*

Palladius-*This Roman historian wrote about a child levitating as seen by his own eyes in the 4th century*

Moslems-*In the twelfth century an Iranian was noted for his frequent flights to the treetops.*

France-   *French historian Louis Jacolliot witnessed 2 levitations in India and recorded them in "Occult Science of India".*

Daniel Douglas Home- *He was probably the most recognized levitator of the Victorian age and was seen by many to float on several occasions during the 1870's.*

Joseph Gianvill- *He wrote about levitation by demons in his book "Saducismus Triumphatus" in 1681.*

Oliver Gilbert Leroy- *He wrote in "La Levitation" stories of 230 different Catholic saints with the ability to levitate.*

Dr. A. Imbert Gourbeyre- *He wrote "La Stigmatisation". It contained over 200 Levitation events as well.*

Colin Evans- *He was well known for his ability to levitate himself and many people saw him accomplish the task in the 1930s.*

There seems to be too many occurrences of this seemingly fanciful act to say that these things never happened. In Tibet, we are told that the monks can issue a sound from drums that allow stones to rise up off the ground and it is the sound or vibration that should be investigated here. While the main element in most of the levitations addressed above was meditation that does not mean that a vibrational element was not introduced by the subconscious. Just like converting a particle into a graviton or a photon, all you must do to levitate something is to convert its matter or the matter of objects around it.

Zero Gravity Lifters

To demonstrate how the PreAztec and PreInca could have levitated once that received the electricity from Egypt, we really only need to look at the dozens of designs of electrostatic lifters being built today as shown in the collage below.

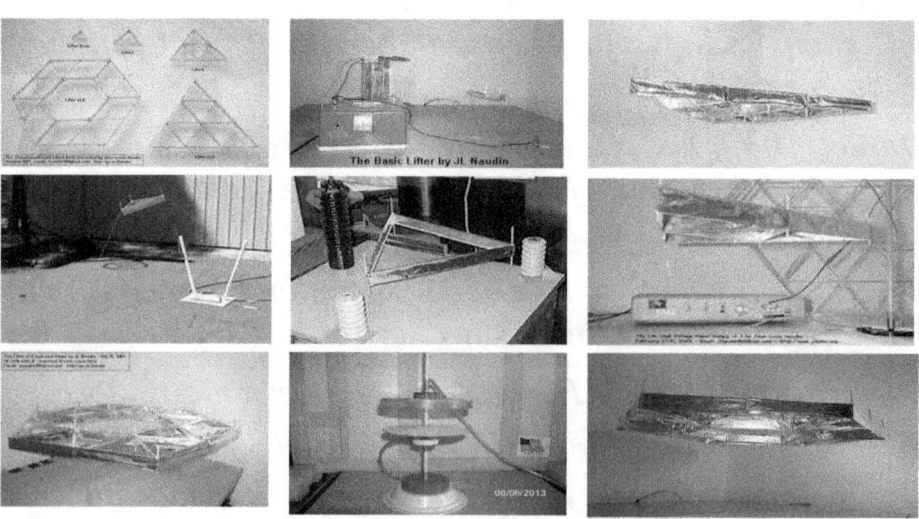

Not only have people re-instituted the art of levitation in one form or another, but also groups around the country are experimenting with new and inventive ways to establish weightless states using some type of pressure built up by high voltage electricity. All you need is about 100,000 volts and make an aluminum foil filed disrupter and up it goes. Here is something interesting. Please notice the one in the middle on the bottom row shaped like a flying saucer.

How in the world could vibrating electricity make matter change its weight, you might ask? Before you read about all this crazy stuff, you wouldn't even believe, matter could change by electrical vibration, but hopefully you are starting to come around. The PreInca possibly were able to generate massive electrical fields around a structure and make blocks lift off the ground just like these things do. Here are a few of the seemingly impossible characteristics established by the lifters:

They use some type of extremely high voltage electricity to float.

The high voltages used many times produce a corona or visible image around the craft.

The vehicles can hover.

They can be extremely quiet.

They can go straight up into the air.

Besides these simple lifters, many more "electro-magnetically driven, gravity elimination devices" have been successfully flown or shown to reduce or eliminate weight. Some have been patented and some stay secret, but these flying objects are not the only things flying around. UFOs of unknown origins have been spotted everywhere around the world. The Flood and the Tower didn't destroy them. Nothing did. They were flying in the past and are even flying today so let's quickly look at some of the facts. The first objection to the lifter evidence above is that they only work with extremely light loads and a reasonable flying machine would have to lift heavy weight. Ed Leedskalnin, John Keely and John Hutchison may give us insight into levitation machines that would lift heavy weights. Each seemed to have re-discovered how vibrational modification could affect mass and gravity. Their efforts may have been enhanced by some of our more well know inventors of the radio, light bulb and telephone. No! I'm not talking about Edison, Bell, and Marconi. I'm talking about the true inventors.

The Restored Vibration Secret

For this section, I'm going to bring out some, not so known, inventors of things vibrational. What I mean by that is that these guys began to re-invent key elements of electromagnetic and magnetic vibration to do work for us. Most of the time, they didn't know what they were doing, but they began to reestablish ancient knowledge of vibration so we need to look at them and what they did to gain perspective. The people I will be examining are the following:

- *Sir Humphrey Davis*
- *Johann Philipp Reis*
- *John Keely*
- *Nathan Stubblefield*
- *Ed Leedskalnin*
- *John Hutchison*
- *Andrea Rossi*

These great men began to open our eyes to vibration as the secret to all things. If you don't recognize their names, shame on you and shame on those who teach us. While we are reading about them, I need to bring out spontaneous human combustion. Sorry, but it's vibrational.

Sir Humphrey Davy

A question comes to mind. Who Invented the Light Bulb? I'll give you a hint it happened in 1809. I think you probably cheated and decided that the section was named Davy, so he must have been the guy. I'm sure you would have said Edison or one of his contemporaries.

All Edison did was to suggest using Tungsten as the emitter so it would last longer. Sir Humphrey Davy, the inventor of the light bulb is shown to the right and this electrical inventing was not his day job. He was a chemist and a great one. He discovered sodium and potassium, calcium, magnesium, boron, iodine, and barium. Then he started getting a little crazy and became well known due to his experiments with the physiological action of some gases like laughing gas (nitrous oxide), to which he became addicted.

179

Just imagine his liking the nitrous stuff so well that he was actually addicted to it. He said that it bestowed all of the benefits of alcohol but was devoid of its flaws. I think he was laughing when he said it.

Miners were getting killed from materials they used in the mines to see so he made them a new light bulb. The bulb used platinum as the emitter rather than tungsten, but it was seventy years before Edison's patent.

As current was passed through platinum, he found out, the platinum began to change into a vibrating mass that vibrated at so a high a frequency that it glowed or had photons being emitted. [His bulb is shown above.]

He made photons with heat. Let's assume our body increases its temperature to protect us from infection. Don't think of it as heat killing bacteria but low frequency photons.

Human Combustion

Here is another thing to consider here. This vibrating photon stuff was created without changing the original material. Just how odd is that? We are told mass can't be created and then we are told that photons sometime have mass and it was created simply by vibrating a mass fast enough. Hopefully you are seeing how all of this stuff is sort of melding together. There is no difference between mass, light, heat, or electricity. It's all the same thing. Speaking of converting body temperature to photons, all these ancient tests about the aura of an angel or person becoming visible to others, should be no surprise. If you can vibrate you body to make it kill bacteria, certainly you could make yourself glow. It's not magic or even odd. I'm sure that if your eye registered slightly lower photonic vibrations, you would see everyone glowing. It certainly would be neat to make yourself into a light bulb and the bacteria would all die in the process. Wait just a minute. That sounds like spontaneous human combustion.

Lightning may not be the worst way to die; what about death by fire? Most people have heard about spontaneous human combustion and immediately ignored it due to its absurdity even though there have been very few explanations presented for some very bizarre occurrences.

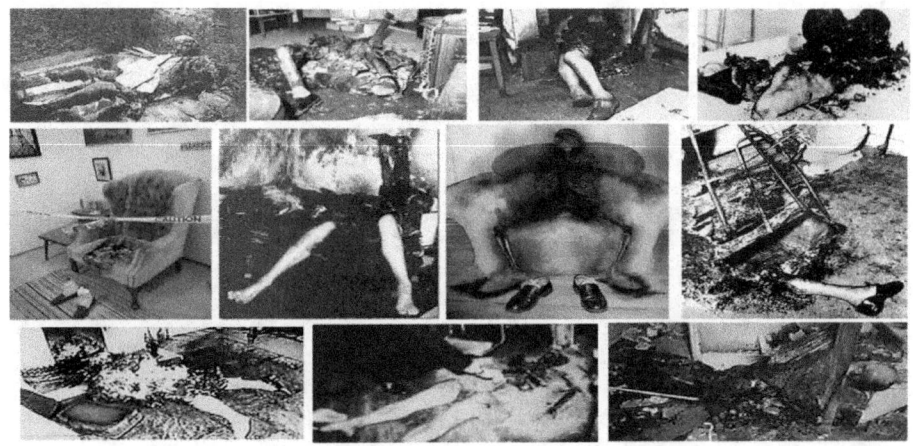

Some had even seen pictures like the ones above that showed the effects of an unexplained phenomenon, but were unmoved. Unfortunately, for Jacqueline Fitzsimons, in 1985 it became too real and there were many witnesses of its reality. She just went to college one day, told her friend she wasn't feeling well and that her back felt extremely hot. All of a sudden, her shirt caught on fire and she screamed for help. Many came to her rescue and put the flames out, but not before her hair was a blaze. After 15 days in intensive care, she died. History and science ignore what has been happening in the past. It is now a little harder to ignore. People can catch on fire by some internal means, so if you get heartburn carry antacids and possibly a fire extinguisher. On the following page is a short list of the more notable instances that have been determined to be spontaneous human combustion in just the past 50 years. This is not an isolated case.

When	What	Where
1951	Mary Reeser burst into flames. Only the corner of the room was burnt.	Florida
1950's	A secretary burst in flames. Her boyfriend was unable to put flames out.	London
1953	Esther Dulin burst into flames. Only she & her chair were consumed.	Los Angeles
1957	Anna Martin burst into flames at home	Penn.
1959	Jack Larber burst into flames in a hospital	Calif.
1959	Billy Peterson burst into flames in a car. The seat was undamaged	Michigan
1964	Fiery death of a man was reportedly like an exploding person	London
1964	Helen Conway's torso was found charred in her home.	Penn.
1964	Olga Worth Stephen burst into flames in her car	Texas
1966	George Turner was found incinerated in a ditch.	Chester
1966	Willem Bruik burst into flames in a car.	Holland
1966	Dr. John Irving Bentley was incinerated at home. His leg was unaffected	Penn.
1967	Robert Bailey burst into flames. Flames were coming from his stomach.	London
1970s	A building Contractor burst into flames while waving to employees.	London
1980	Mr. Blackwood was incinerated but rubber on his walker was not burned.	Wales
1980	Jenna Winchester burst into flames while sitting in her car.	Florida
1982	A woman burst into flames on the street.	Chicago
1982	Jean Saffin burst into flames in her kitchen. Her father remained helpless.	London
1985	Jouqueline Simmons burst into flames at school.	Cheshire
1989	A 27-year-old engineer burst into flames. His stomach and belly were carbonized.	Hungary
1997	John Oconner burst into flames at home. Only his head, upper torso and feet remained.	Ireland
1998	Agnes Philips burst into flames in her car. Flames came from her chest.	Australia
2003	Alexei Rusnac's head was burnt while his clothes were not burned according to "Ananova"	Rumania

What the list shows is that spontaneous "Spontaneous Human Combustion" occurrences are all over the place, and people spontaneously combust just about every year. Some years more incidences seem to happen than others and the following general information concerning the symptoms of the mystery are described below.

- *Although it doesn't appear that way from my short list, almost eighty percent of the victims are female.*

- *Most of the victims were overweight and/or alcoholics.*

- *The body is very badly burned, but the room the body is found is in pretty much intact except for a fine layer of soot.*

- *Yellow, foul smelling oil usually surrounds the body.*

- *The torso, including the chest, abdomen and hips tend to be totally consumed, sparing portions of the extremities and the head - the clothing can also be intact.*

- *Generally, victims were on their own. Generally, no shouts or screams could be heard. This was not the case with Jacqueline.*

- *The victim had usually been drinking heavily prior to the death, but that isn't always the case.*

The list doesn't help very much as we try to find a cure for this affliction, but hopefully you are beginning to believe that there are things about our human bodies that we have little knowledge about and I don't only mean that one day we may blow up.

Spontaneous Feline Combustion

Of course, humans are the only animals that should worry about this phenomenon. No one will probably know about most of the explosions, but Peppi the cat's flames were seen quite clearly. The day was the 28th of November 1986. He was an 8-year-old house cat and just sleeping on a chair at the Anmer Lodge in Stanmore London. Witnesses said that there was a terrific bang followed by a flash going a few feet in the air. Peppi was enveloped in a blue flame similar to other spontaneous combustion victims and essentially disintegrated. Peppi was another victim and still there is no answer to the mystery.

Combustion Discussion

As with most everything else, I have a vibrational theory about this strange anomaly. All the other explanations seem to fall away from one incident or another, but there may be a broad theory that allows for and explains how this terrible thing could happen. The Tamashii atomic model helps explain what may be happening. With this general concept in mind let me say that if the right electromagnetic conditions surrounded the unfortunate "combustion" victims, some of the molecules in their bodies could have easily been converted into some type of combustible material. That doesn't mean that when you get heartburn you should go for a fire extinguisher, but it does mean that there is a logical explanation for things that may initially appear to be odd. New researchers may be getting closer to the answers of these exploding people by making bowling balls fly in the air on their own. I know that sounds like an impossible thing to happen, but that doesn't mean that it didn't happen. We will find later that a Canadian named John Hutchison is causing many "impossible things" to happen by using extremely high frequency signal bombardments. After many successes and more failures, he can tell you that it is extremely difficult to get the right combination of fields to cause changes in materials, but when they change, the effect is dramatic and you don't need to use ATOMIC FISION or FUSION or

184

anything else that could be really bad. Human combustion is one of the dramatic changes that can occur from a changing FREQUENCY. There might, however, be a reason why many of the instances involved alcohol. The possible reason is that alcohol causes desensitization of nerves in our body which allows for misfired electrical signals. Information paths are, then, disrupted more by the chemical imbalance at the nerve endings. While the misfired signals would not probably cause the reaction on their own, they may help in the modification of the internal molecular makeup that results in a "self-induced" explosion. Just like extreme anger seems to bring our bodies to an extremely heated condition, this action probably requires many individually unimportant events to occur simultaneously.

I know this explanation doesn't help us control, or know the specifics about why each of the events occurred, but it does provide a link between spontaneous human combustion and reality. As electromagnetic beat frequencies are funneled or even accidentally manufactured, very strange anomalies like levitation, body combustion, healing someone with the mind, and even out of body visions may result. Each of these anomalies have had many, many, witnessed occurrences. Just because we don't quite understand the physics doesn't mean that there is not a real physical law that allows for these strange things to be part of our normal life. For a short time, conversion of materials from noncombustible to highly flammable occurred inside Jacqueline Simmons and Peppi the cat so don't say it is too farfetched to happen. If they heard you saying something like that they would disagree strongly. Speaking of hearing, let's get back to the invention of the telephone.

Johann Philipp Reis

Someone might ask you, "Who Invented the Telephone?" You might say Bell or any of his contemporaries of 1890, but you really have to go back a little further when Alexander Graham Bell was a child. The inventor was actually a German physicist named **John Reis**. In fact, he invented the telephone in 1861.

 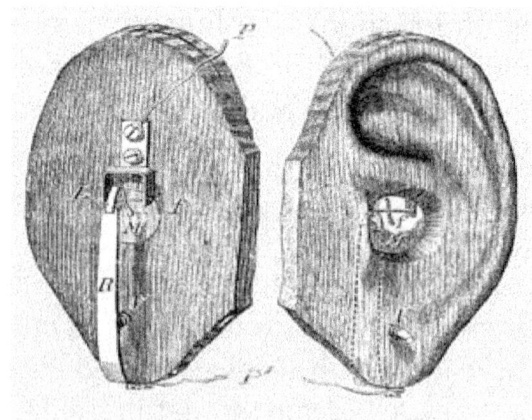

This was many years before Bell became famous with his copy. About the only change made by Bell was the use of a different voice diaphragm to increase the clarity. Our history books praise Bell. Probably German history books tell a different tale. The sound quality wasn't very good, but the device worked, just the same. John Reis had rediscovered a principle of vibrating electricity to produce vibrational magnetism used to regenerate sound along a wire. It was a marvelous invention.

As a matter of curiosity try to guess what the first words that were successfully transmitted. You guessed it-- *"the horse eats no cucumber salad"*. Not knowing what to call his new invention, he said it was a **"Telephon"**.

For a receiver, Reis built a basic receiver and attached it to part of a violin to act as a sounding board. Here's the funny part. His transmitter was carved out of a piece of a beer barrel and he made it in the shape of an ear. [as shown on the preceding page]. "I need a diaphragm", he must have exclaimed and he found a sausage skin to form it. Then he added a tiny strip of platinum and glued it to the diaphragm as one electrical contact. The other wire was fixed to an adjusting screw. You could hold the ear up to your ear and hear stuff. This strange invention is still preserved in the Patent Museum in Berlin.

Reis actually got his inspiration from a paper written by Bourseal in 1854. In this paper he said, "Speak against one diaphragm and let each vibration "make or break the electric contact. The electric pulsations thereby produced will set the other diaphragm working, and it then reproduces the transmitted sound." The make or break thing wasn't a good idea, but it got Reis going. His diaphragm actually drove a thin rod to varying depths in an electric coil to make a vibrating magnetic field. Reis died two years before Bell received his patent. He was only 40, and he never did get around to seeking a patent for his device.

The drawing below is the 3rd version of the Telephon. The ear had been replaced, but you could still hear things.

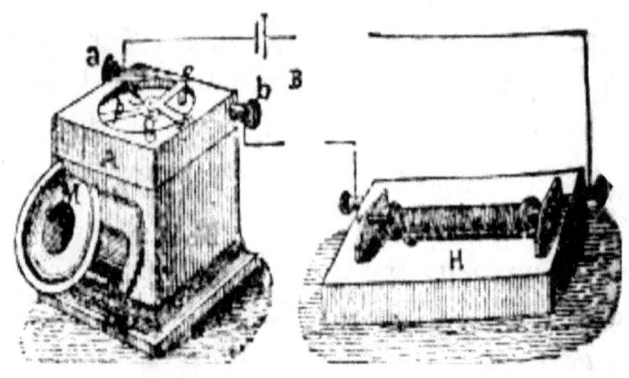

Reis had converted vibrating electricity into magnetism and the magnetic field vibration produced vibrating air or sound.

John Keely

Mr. Keely is gone now, but during the 19th century, his discoveries spawned new light into thoughts of free energy, vibration, and levitation. Unfortunately, when he passed away in 1898, all his information was lost with him.

He, evidently, had re-discovered what the ancients had used in the past and he did it with a variation of the Tamashii model of the atom. I'm not going to reconstruct this new way of looking at matter, but the statements below can be easily derived from this very important study.

General Tamashii Model -The subdivisional parts of atoms react with each other in proportion to vibrational differences.

Its Corollary-Modification of these elemental characteristics including visibility, gravity, and electronegativity can be modified by affecting the vibrational differences.

Here are a few things he had to say and things he demonstrated with his discoveries. See how similar they are to the new science being introduced today.

Keely's Law of Vibrating Subatomics- *Atoms are capable of vibrating within themselves at a pitch inversely as the local coefficient of Gravity, and as the atomic volume, directly as the atomic weight, producing the "creative force", whose transmissive force is propagated through subatomic solids, liquids, and gases, producing induction and the static effect of magnetism upon other atoms of attraction or repulsion, according to the law of harmonic attraction.*

Keely's Law of Oscillating Subatomic Particles-*Subatomic particles, oscillating at a uniform pitch produce the "creative force", whose transmissive form, "Gravism", is propagated through more rarefied media, producing the static effect upon all other sub atomic particles, denominated as Gravity.*

Keely's Vibration Definition-*Life in its manifestations is vibration. Electricity is vibration. But vibration that is creative is one thing. Vibration that is destructive is another. Yet they may be from the same source. As in the electrical forces in the form or nature prepared even for use in the body."*

Keely's Inventions-With his science of vibration, Keely invented many devices. Here is a list of the better-known experimental results.

- *Exploding water and producing 29,000 PSI, by use of some type of sound wave.*

- *Disintegrating quartz crystals by using acoustics.*

- *Producing rotation by compound sound waves (patented in modern times by Panasonic as ultrasonic motors) an engine was reportedly driven by the flow of Aether into its components.*

- *Tapping into what he called "Aether flows" to run his engines.*

- *Producing a glowing blue light in water using acoustics (now rediscovered as 'sonoluminescence').*

- *Producing a compound "frequency driven" motor that ran from many frequencies.*

- *Demonstrating a pneumatic cannon powered by release and instant expansion of some kind of bizarre plasma vapor.*

- *Demonstrating an acoustic based flying machine that levitated and propelled itself in the presence of government witnesses.*

- *Demonstrating an acoustic microscope capable of viewing into the molecular and atomic interstices of matter.*

- *Demonstrating a globe which could be made to rotate with no outside source of power as a demonstration of the Aetheric flows into matter.*

One of his famous machines is shown beside his preceding image. I have no idea which one, but Keely certainly, was close to funding out how vibration changed matter.

Nathan Stubblefield

Who Made the First Radio? I guess you know it wasn't Marconi, because I purposefully picked things that have been changed by history books. In this case the inventor was Nathan Stubblefield and it happened in 1892. He demonstrated his device over a period of 10 years to many people, but unfortunately his patent "00600457" didn't keep others from stealing his invention. Many years later Marconi became our hero. Nathan's radio is shown to the right.

As the true father of the Radio, Nathan said. "I have perfected now the greatest invention the world has ever known. I've taken light from the air and the earth as I did with sound ... I want you to know about making a whole hillside blossom with light..." After that, he locked himself in his shack and starved himself to death. Nathan is actually known [or should be known] for 3 major vibrational inventions.

The radio was the last. Prior to that one he invented the mobile phone. No, he didn't carve it out to look like an ear. His phone used magnetic transfer instead of electric. Let me explain. His phone was mostly made from a huge coil of wire. If another larger coil of wire was around the phone area, coming within ½ mile of the source would allow the vibrating magnetic field of the transmitter to cause an equivalent vibrating magnetic field on the receiver.

The vibrating rate was audible. His receiver and transmitter base are shown above. That little bit of discovering wasn't the only thing he discovered. In fact, his first discovery was that the earth magnetic fields vibrate which in turn can cause galvanic action between metal bars pushed in the ground.

By coiling dissimilar metal coils together, he was able to remove energy from the ground because of this effect.

Stubblefield had converted vibrating magnetism into vibrating electric fields that could be converted into sound.

Ed Leedskalnin

The guy, shown on the right, immigrated to the United States from Latvia in 1923 and quickly contracted the deadly Tuberculosis disease. Somehow, he cured himself. That wasn't the amazing thing to consider here. He also "harnessed gravity itself" it would seem. He picked up 1100 tons of coral rock, carved the stones, and built a castle "alone", and with no help whatsoever. In 1952, this apparent magician died, but before he passed away he told many about his ability to control gravity. After his death, a strange machine was found at his castle.

 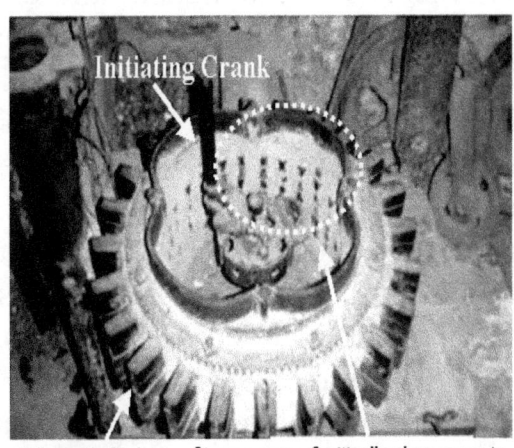

Initiating Crank

Magnetic Stator   Strange array of critically place magnets

As shown above, the machine had many magnets embedded in rock and others that were positioned by turning a crank on the top. While the portion of the machine seems to be incomplete, we can image this machine to be somewhat similar to the no-energy machines being produced around the world today, but this one, somehow affected gravity. As the machine was turned, it would have generated some frequency

195

associated with the times that the rotor magnets passed near the embedded ones.

If the generated vibrational components could have been high enough, the vary essence of the nearby materials he was trying to affect would have changed. Evidently the bolders he moved turned into something quite different than what they are today so that he could easily place them in their locations. They evidently loss most of their mass or the gravity component of the mass while under the influence of Ed's machine.

Mr. Leedskalnin insisted that all matter was **not** made of atoms but, instead, consisted of individual magnets and it is the movement of these magnets within materials [Vibrational element] and through space that produces measurable phenomena, i.e., magnetism and electricity. He also claimed to know how the Great Pyramid was built, and to prove it he moved a 30-ton and other monolithic blocks of coral to build his castle. He inferred that out-of-phase gravity waves can be created in such a manner to neutralize in-phase gravity waves. While most of my discussions is about changing particles by changing the vibrational component remember that the gravity component appears to be a perpendicularly phased vibrational component that works in tandem with the normal vibrational component associated with the characteristics of matter. Ed understood what ancient humans had known during the very ancient times and we are just now relearning.

No one really knows what Ed did, but you can be sure it had something to do with vibrating magnetic fields converting matter.

John Hutchison

Luckily, Mr. Leedskalnin wasn't the last man to rediscover levitation, disintegration and free energy by vibration. We cannot talk about modern levitation without discussing John Hutchison.

He currently lives in Canada and has been conducting some amazing, well witnessed experiments since 1979. He will soon find out what produces what we call the Hutchison effect. Right now, let me just list the elements known so far. He has demonstrated all of them in the presence of a strange field of electromagnetic waves.

- ***Objects*** *became temporarily invisible*

- ***Heavy objects*** *can levitate [even a bowling ball lifted into the air]*

- ***Things can pass*** *through each other without deforming the material [See below]*

- ***As they pass*** *through each other there is no apparent change in either component's physical characteristics.*

- ***Sometimes metals*** *can become like jelly [see below]*

- ***Sometimes metals melt*** *without heating*

Hopefully, your interest has peaked. The inventor uses multiple radio transmitters and high voltage Tesla coils concentrated at a single location to produce "something" and when it is produced, the above things occur to objects in this "Field". If you noticed a link between the multiple radio transmitters and the descriptions of the "Tamashii model of atomic structure" and the use of vibrations to transform elements, I think you are on the right track here.

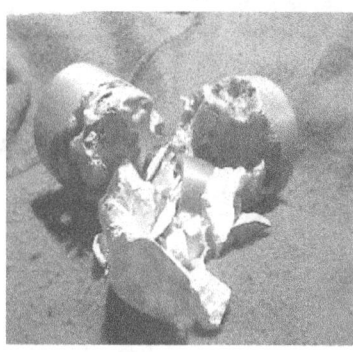 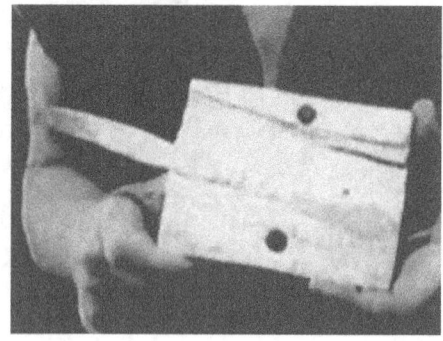

John Hutchison has accidentally found the correct vibrations for particular elements to make them appear invisible to one another so that one can pass through another. The picture above to the right shows how a simple butter knife was pushed through a piece of metal without anyone's help. Now cut in half, the butter knife has become part of the metal. He also has made objects appear to be gravitationally invisible to the earth to allow levitation. During some of his experiments, vibrations mutated the materials to appear to be melted or jelly-like. [See the metal bar above to the left.] Sometimes metals would appear to melt but surrounding wooden objects would not get hot as only the metal structure changed in the field. The effect was as if the objects were changing from this universe to another. The stuff John was finding out was incredible, but someone got scared. In 2006, the Canadian Government went up to John's home and confiscated all of his vibration altering equipment.

While "vibrationally" we may be able to define and even modify a particle, if we are to understand the universe, we must expand our awareness to the 12-dimensions that make up the vibrational entities.

Rossi and LENR

If we want to talk about modifying matter with vibration we have to discuss something called LENR, <u>L</u>ow <u>E</u>nergy <u>N</u>uclear <u>R</u>eactions, also known as CANR, <u>C</u>hemically <u>A</u>ssisted <u>N</u>uclear <u>R</u>eactions. The first commercialization of this type of modification of matter was accomplished by an Italian inventor named Andrea Rossi shown below. By slamming Hydrogen ions into nickel, with the right vibrational coding allows the nickel to convert itself into copper and release heat used for electricity. Actually, they haven't tied the vibrational mode to the process yet, but they will over time.

It seems everywhere you turn another scientist is trying to figure out what is happening as Nickel is magically turned into copper while producing massive amounts of heat. One list I saw had over 3,500 journal papers, news articles, and books about LENR. This is the conversion of one material to another by presenting a vibrational difference, near the proton head so that the nuclear force is modified and the material is changed which can used to make vast amounts of electricity without a pyramid.

The plan is to eliminate electrical grids and have each house power itself and have automobiles powered by this matter transfer system. A typical home system is shown next upper left followed by a larger factory LENR reactor.

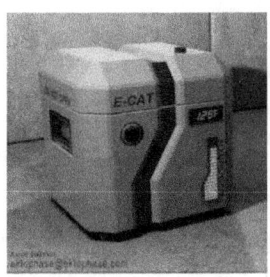

A number of automobiles to use LENR are under design as is a strange jet and something that is in between being considered by NASA.

Not to stop there, a speed boat will use one of the reactors and even lightweight motor bikes all shown.

Invisibility and Alchemy

Beside changing matter for electricity, people have been trying for years to replicate ancient capabilities in invisibility, alchemy and many other things. Sometimes we have gotten pretty close. One time might have been called the Philadelphia experiment.

Philadelphia Experiment

If we could better understand the Tamashii Atomic model, Einstein's unified particle theory, Tesla's Ground current phenomenon, the Acoustic Laws of Keely, or the Hutchison Effect that have been touched on throughout this work, maybe we could better understand what might have happened in Philadelphia on board the USS Eldridge during World War II as shown next.

Most of the reports of that experiment can't be verified and initially seemed fanciful. The reports of people become fused with the bulkheads, and the ship disappearing and reappearing at a different location all seem more like science fiction, but maybe there is more to the stories that we can begin to understand. I'm not getting into this area as many already have heard about the outcome of this experiment except to say that people should not have been inserted into the experiment. People were supposedly found merged with bulkheads and the survivors went mad. Rather than suppressing these experiments, they should be heralded. If we can understand these oddball physical anomalies, maybe we can better understand something we call Alchemy.

Modern Alchemy

This section is on successful transmutational experiments done in the 20th century; well before the LENR devices. I know alchemy is a bad word and instills an image of sorcery, but in the 20th Century, it is almost commonplace to transmutate elements into other elements. The following list describes the successes that have been reported. Today most are done with the "Atom Smashers" so it is still too expensive to make gold for most of us. As can be seen from the table below, some are beginning to use high voltage and frequency methods which go along with the Tamashii model. These methods don't take up nearly as must space as an atomic accelerator.

Date	Researcher	Method	Initial Mat'l	#	End	#
1920s	Franz Tausend	A	Mercury	80	Gold	79
1923	A. Miethe	B	Mercury	80	Gold	79
1924	Hantaro Nagaoka	B	Mercury	80	Gold	79
1924	Smits & Karssen	B	Lead	82	Mercury	80
1927	Walter Russell	B	Oxygen	8	Nitrogen	7
1935	Lord Rutherford	C	Nitrogen	7	Oxygen	8
1936	Various	C	Platinum	78	Gold	79
1938	Hahn & Strassman	C	Uranium	92	Ba + Kr	56+36
1939	Various	C	Thorium	90	Radium	88
1960s	Jnana & Caro	A	Lead	82	Gold	79
1972	Soviet Physicists	C	Lead	82	Gold	79
1980	Glenn Seaborg,	C	Lead	82	Gold	79
1986	Various	C	Mercury	80	Gold	79

With this list, there is a strong echo of the "Book of Secrets" warning. According to that ancient work, God warned that-- "Man can change elements just like he can change animals, but he cannot understand the outcome of his actions." By the way; the **A** means **Harmonic Alchemy** [not really frequency based]—**B** means **High voltage/high frequency distillate**— **C** means **Nuclear Accelerator Bombardment**. Notice that the most converted element is Gold. I guess there is a flare for sorcery in our modern-day scientists.

Beyond

Some of you are wondering why I have gone over these bizarre quasi-scientific experiments using anomalous methodologies and identified questionable applications of possible uses of vibrational knowledge from thousands of years ago. The answer is simple. No one else seems to be providing this information to those who might be able to use the information the most. That group is our children. They may be able to better understand these "phenomenon" if they can build on the previous experimentation. Hopefully you are considering the possibility that the ancient texts and the other evidence of almost magical capabilities were not lies or fairytales. Our new understanding of matter in general and how things can completely change characteristics by simply changing a vibrational element of a system is now allowing us to accept ancient events that previously separated religious accounts and scientific understanding. Heaven not only is a possibility but a necessity and the probability of angels existing now has become reasonable. Lastly, a photon can be a particle sometimes and a wave [or vibration] sometimes.

I need to go a little further to allow a more complete vision of this new concept of reality. As I mentioned before "string theorists" tell us that 6 of the 12 dimensions needed to create our universe are "compactified". I explained to you, when I stated it, that their concept simply makes no sense. If there are 12 dimensions needed in the universe there must be 12 dimensions working in this universe and that is what we are getting into in the next book of this series.

Conclusions

1. Early attempts at defining the unified particle always ended in failure. Bayrons [electrons] were broken down into Borons [the smallest particle containing all elements of existence].

2. Unfortunately, the Borons could be broken down into FERMIONS which had component parts of particles but not everything. The one usually used as an example is the graviton which has gravity, but no mass to support the gravity.

3. Unfortunately, there were many types of fermions so something else made them.

4. All matter is made up of vibrational fields instead of particles.

5. Vibration levels determine the character of a particle. As a photon vibrates faster, its characteristics chance from visible light all the way to massively dangerous COSMIC rays that can go right through your body.

6. If the electromagnetic vibrational pattern is excited faster, the photon becomes a particle.

7. As the vibration still increases in speed, the particles get larger and larger. Gold, for instance is vibrating much faster than helium.

8. Just like noise reducing headphones make sounds disappear, vibrating particles become invisible to each other as they become in 180 degrees out of phase.

9. Beyond that, everything in the universe can be broken down into vibrations. This includes not only the things we can see like planets and tiny little pebbles, but also electricity, magnetism, Photons, Nucleons, the spirit, the soul and even life.

10. You should have also gotten a better understanding about how the Christian religion fits with all of this stuff. How God incarnate came to reestablish the link between Heaven and Earth. How demons could be understood. How reincarnation could be addressed. How the final rapture of the Church or the 2nd coming of God incarnate could fit into a world not of 4 dimensions as you once thought but in a world of 12 dimensions.

11. This was not a religion book, but religion must coincide with our mathematical models or we have one of 2 problems. Either the science is wrong or the religion is wrong or incomplete. If the model is correct, religion and Science will agree.

Hopefully this book has been useful to you when trying to interpret what otherwise was considered anomalous in our world. If too many things don't fit, you must try to change your definitions. This is a new definition.

The End

About the Author

Steve Preston is a long lime author of scientific, esoteric facts. His series on the creation of mankind is shown below. The series focuses on the painful truths rather than whitewashed details that make us comfortable. If you are interested in the truth instead of comfort, please continue to read and, while you are at it, review other works by Mr. Preston as shown below.

Development of Mankind

The First Creation of Man-book 1 History of mankind
The Second Creation of Man-book 2 History of mankind
The Creation of Adam and Eve-book 3 History of mankind
The Antediluvian War Years-book 4 History of mankind
Man After The Flood-book 5 History of mankind
A Closer Look at Ancient History-book 6 History of mankind
A New View of Modern History-book 7 History of mankind
The Twentieth Century and Beyond- Book 8 History of Mankind

Bible History, Correction, and Analysis

Abraham to Moses-First part of the Bible
Adam's First Wife-Story of Lilith
Adam to Abraham- Second Part of the Bible
Closer Look At Genesis- 200 ancient text confirm Genesis
Exploring Exodus- Reviewing the Details of "Exodus"

Errors in Understanding- Interpretations of the Bible
Expanded Genesis- Apocrypha and other Jewish texts
Exploring Genesis- Reviewing the details of "Genesis'
Incarnations of God- How often did God become Incarnated?
History Confirmed By The Bible- Science confirms the Bible
Moses Saved Egypt- How the Jews eliminated the Hyksos
Moses to Jesus- Third part of the Bible Series
Mysteries of the Exodus- Proofs of the Exodus
New look at the Bible- Questions in Interpretation
Old Testament Used By Jesus- Ancient Jewish texts
Understanding the New Testament-4th part of the Bible Series
Why the King James Bible Failed- Issues with KJB

Ancient Technology and Life

Anakim Gods- History of the Ancient Giant/gods
Ancient History of Flying- Ancient flying
Kingdoms Before the Flood- Pleistocene humans
Living on Venus- Venus before the Pleistocene Extinction
Martians- Ancient Life on Mars
Mysterious Pyramids- Who made the Pyramids?
Victory of the Earth- History of our Earth
Not from Space- UFOs are not from space.
Amazing Technology- Descriptions of prehistoric capabilities

Ancient and Modern War

America's Civil War Lie- Truth about the Civil War years
Behind the Tower of Babel- Story of the Bharata War
Driven Underground- Fear in the Bharata War
Four Armageddons- The 4 major wars that destroyed mankind
Six Deaths of Man- Destructions of mankind
World War Before- The Pleistocene War
World War with Heaven- The Angel and Anak War
World War Zero-The Bharata War

209

When Giants Ruled the Earth- History of the Titan Giants
Sex Crazed Angels- What caused the Heaven War?

Current Events and Fears

Allah' God of the Moon- Terror of Muslims
American School Disaster- fear in our country
Can We Save America? - Fear in the USA
Scythians Conquer Ireland- A History of Ireland
Fast History of MILES Training- Laser based Army training
Great American Quiz- Unusual details of American History
Make Your Own Global Warming
Truth About Phoenicia- The Evidence -First in America
Monsters are Alive- Post Pleistocene Monsters
Promote the General Welfare- Fear in USA
Our Very Odd Presidents- President review
Terror of Global Warming- Fake issue uncovered
The Antichrist- Many demonic possessed rulers
The Bad Side of Lincoln- Negative side of a great man
The Devil- Of Demons and their master
Vampires among Us- How Demons and Vampires are similar
Humans on Display- Slavery and Human Zoos

New Look at Physics

Amazing Technology- Pleistocene Technology
Anthropic Reality- We control our Reality
Consensus Science- Fake Science
Complex Earth- Truth behind Earth's development
Is Time Travel Possible? Science of Time Travel
Retiming the Earth- Eliminate of Nuclear Decay Errors
Releasing Your Consciousness- Beyond our SELF
Slip Through a Wall- How to walk through solids
Our 12-Dimensional Universe- New science of our Universe
Mystery of Photons and Light- Science of Photons

Of Heaven and Hell- scientific descriptions
Meaning of Life and Light *- Detains of New Science*
Vibrational Matter *- New Science of Quantum Fluctuations*

New Look at Biology

DNA of Our Ancestors- Tracing DNA of ancient man
God Didn't Make The Ape- New science on ape Evolution
Lizard People- Mutated People of the Bharata War
Creation and Death of Dinosaurs- Why Dinosaurs died
Races of Men- Tracing DNA of Humans
Tracing Cro-Magnon to Jesus-
Self, Soul, Spirit- Three components of Life
Self-Virtualization- New science of reality
True Happiness- Self Actualism and Beyond
Life Resonance- Unusual capabilities of men
Awaken the Departed- We can talk to the Dead
Biophotonics and Healing- How Photonics used in medicine

www.ingramcontent.com/pod-product-compliance
Lightning Source LLC
Chambersburg PA
CBHW051645170526
45167CB00001B/337